QUANTUM-DOT BASED LIGHT-EMITTING DIODES

Edited by **Morteza Sasani Ghamsari**

Quantum-dot Based Light-emitting Diodes
http://dx.doi.org/10.5772/65178
Edited by Morteza Sasani Ghamsari

Contributors

Zhifeng Shi, Xinjian Li, Chongxin Shan, Anca Armăşelu, Namig Hasanov, Elena Kolobkova, Nikolay Nikonorov, Neda Heydari, Seyed Mohammad Bagher Ghorashi Ghorashi, Hyung-Ho Park, Wooje Han, Fanyao Qu, Morteza Sasani Ghamsari, Alice Bernard

Published by InTech
Janeza Trdine 9, 51000 Rijeka, Croatia

Publishing Process Manager Dajana Pemac
Technical Editor SPi Global
Cover InTech Design team

First published October, 2017
Printed in Croatia
Legal deposit, Croatia: National and University Library in Zagreb

Additional hard copies can be obtained from orders@intechopen.com

Quantum-dot Based Light-emitting Diodes, Edited by Morteza Sasani Ghamsari
p. cm.
Print ISBN 978-953-51-3575-3
Online ISBN 978-953-51-3576-0

For EU product safety concerns:
IN TECH d.o.o., Prolaz Marije Krucifikse Kozulić 3, 51000 Rijeka, Croatia,
info@intechopen.com or visit our website at intechopen.com.

Meet the editor

Dr. Morteza Sasani Ghamsari is a senior researcher in the Photonics and Quantum Technologies Research School of the Iranian Nuclear Science and Technology Research Institute. He has respectively received his BSc and MSc degrees from Sharif University of Technology in Iran and PhD degree from the National University of Malaysia (UKM). His research focused on the preparation and characterization of solid-state laser crystals, scintillators, and nonlinear optical materials. His recent interests lie also in nanophotonics including metamaterials, quantum dots, plasmonic nanomaterials, and photonic nanodevices. He is an editorial board member and a reviewer of different international journals, and has collaborated with some local and international academics/researchers on graduation of postgraduate students and research projects. He has published 3 book chapters and over 80 articles in scientific journals as well as in reviewed conference proceedings.

Contents

Preface

This book was motivated by the desire others and we have had to further the evolution of the light emitting diode (LED) technology. It was also aimed to provide a review of recent developments in the quantum dot (QD)-based organic light emitting diode (OLED) technology and its application.

In the beginning, the book provides a general introduction and then gives some information on the light emitting diode (LED) technology. In further chapters, the technological aspects of the light emitting diode (LED) manufacturing were covered by the authors. Subsequently, the book tried to explore the challenges of quantum dot–based light emitting diode (QD-LED) technology. Consequently, the recent progress in fabrication techniques and construction of QD-LEDs was noticed in one of the chapters, and the other one focuses on the recent development in the application of QD-LEDs.

The embedded CdS(Se) and PbS(Se) quantum dots with high room-temperature quantum efficiency in the fluorine phosphate glass matrix were evaluated in one of the chapters. The authors of this chapter focus on how these fluorine phosphate glasses were used in light emitting diodes (LEDs) and present the main advantages of applying these glasses in the LED structure.

The recent progress on newly emerging perovskite light-emitting diodes based on organic-inorganic hybrid perovskites ($CH_3NH_3PbX_3$, X=Cl, Br, I) and inorganic perovskite cesium lead halide ($CePbX_3$, X=Cl, Br, I) quantum dots was also considered in another chapter of this book. In the other chapter, the critical role of quantum dots to achieve full color conversion or white light generation in solid-state light sources was described in detail. This book provides a complete picture of the field for advanced undergraduates, postgraduates, and researchers.

Morteza Sasani Ghamsari
Photonics and Quantum Technologies Research School
NSTRI, Tehran, Iran

Section 1

Organic Light Emitting Diodes

Chapter 1

Introductory Chapter: Quantum-Dots Based Organic Light-Emitting Diodes - The State-of-the-Art

Morteza Sasani Ghamsari

Additional information is available at the end of the chapter

http://dx.doi.org/10.5772/intechopen.69744

1. Promising QD-OLED technology

Organic light-emitting diode (OLED) is a light source in which a thin layer of organic material placed between two conductors that can emit light of specific color by applying electrical current. To accelerate the commercialization of organic light-emitting diodes (OLEDs) have enhanced due to their capabilities as new generation displays and lighting sources. In comparison with other display technology, OLED displays are thin, efficient, flexible and transparent. In addition, they have better contrast, higher brightness, fuller viewing angle, a wider color range, much faster refresh rates and consuming lower power, thinner, very durable and also they can operate in a broader temperature range. But they have some disadvantages too. The cost of OLED displays is still too high. They have limited lifespan and suffer from permanent image retention. If these limitations can be overcome, OLED displays can find faster growth **opportunities** in specific **display technologies**.

In recent years, the display industry is rapidly interested in the world to combine semiconductor nanocrystals or quantum dots (QDs) with organic light-emitting diode. Semiconductor nanocrystals or quantum dots (QDs) have considerable potential in assisting OLED displays and lighting to overcome their technological barriers. They are a type of nanomaterials which exhibit good optical characteristic. One of the important properties of QDs is that the absorption and photoluminescence (PL) spectra can be adjusted by their size. Consequently, the pure and tunable spectra of QDs can be conventionally obtained. Therefore, QDs can be used in photovoltaic devices, sensors and light-emitting diodes (LEDs). The smaller dimension of the QD results in shorter wavelength light emission and longer wavelength of light emitted will cause by bigger QDs. Therefore, it was believed that the next generation of OLED can be introduced by employment of this type of photonics material in manufacturing process of OLED.

Researchers all around the world try to improve this field and they found that quantum dots (QDs) can be helpful. Excellent color-rendering properties and high luminous efficiency (LE) of OLED are obtained when the combination of colloidal QDs with light-emitting diodes (LEDs) was done. QD-LEDs are the important part of the next generation of solid-state lighting and display technologies due to their great color saturation features, tunable wavelength and narrow full-width at half-maximum (FWHM). Nevertheless many developments have recently been made in display and lighting board technology and this ability was specially developed to build a very thin flexible and transparent displays; scientists have found that using some extra thin layers between two conductors and choosing appropriate materials can be helpful for better performance of QD-LEDs [1, 2]. High brightness, flexibility, efficiency with long lifetime and low processing cost of QD-LEDs makes them different from LCDs, OLEDs and plasma displays.

QD-LEDs can produce dark blacks and whiter whites and create higher brightness than OLEDs. In addition, a wider and more true-to-life color palette will be possible in QD-LEDs than in OLEDs. Because, an improvement in operation lifetime of OLED devices can be achieved by using quantum dots. Finally yet importantly, QD-LEDs are cheaper than OLEDs. Nowadays, researches try to improve efficiency of passing light through the quantum dot crystals, which can help the QD-LEDs performance [3]. QD-LEDs have some unique features such as better color accuracy, higher brightness, more stable performance and lower cost. Also, they are solution processable and suitable for wet processing techniques.

Colloidal nanocrystal-based **LEDs** can be classified into two main categories. In the first type of configuration, QD-doped material applied over the emitter chip and optically excited by a conventional epitaxial blue or UV LED. In the second category, the QDs themselves are directly excited by the passing of current through a QD-containing film. In QD-OLEDs, a layer of quantum dots is placed between layers of electron-transporting and hole-transporting organic materials. By applying the electric field, electrons and holes move into the quantum dot layer, recombine in QDs and emitting photons. The full width at half the maximum value of the QDs shows the width of the spectrum of emitted photons. In that case, a thin emissive layer is sandwiched between a hole-transport layer (HTL) and an electron-transport layer (ETL) to capture electrons and holes in the small region. Organic materials are used as the ETLs and HTLs, until now. The electron-hole recombination generally occurs near the cathode because organic electroluminescent material transport holes faster than electrons. Therefore, the exciton produced will be quenched. In order to prevent this event, a hole-blocking layer will be used.

Author details

Morteza Sasani Ghamsari

Address all correspondence to: msghamsari@yahoo.com

Nuclear Science and Technology Research Institute, Tehran, Iran

References

[1] Li Z. Enhanced performance of quantum dots light-emitting diodes: The case of Al_2O_3 electron blocking layer. Vacuum. 2017;**137**:38-41

[2] Kang Y, et al. Quantum dots for wide color gamut displays from photoluminescence to electroluminescence. Nanoscale Research Letters. 2017;**12**:154

[3] Li Z, et al. Efficient and long-life green light-emitting diodes comprising tridentate thiol capped quantum dots. Laser & Photonics Reviews. 2017;**11**:1600227

Chapter 2

Recent Developments in Applications of Quantum-Dot Based Light-Emitting Diodes

Anca Armăşelu

Additional information is available at the end of the chapter

http://dx.doi.org/10.5772/intechopen.69177

Abstract

Quantum dot-based light-emitting diodes (QD-LEDs) represent a form of light-emitting technology and are regarded like a next generation of display technology after the organic light-emitting diodes (OLEDs) display. QD-LEDs are different from liquid crystal displays (LCDs), OLEDs, and plasma displays due to the fact that QD-LEDs present an ideal blend of high brightness, efficiency with long lifetime, flexibility, and low-processing cost of organic LEDs. So, QD-LEDs show theoretical performance limits which surpass all other display technologies. The goal of this chapter is, firstly, to provide a historical prospective study of QD-LEDs applications in display and lighting technologies, secondly, to present the most recent improvements in this field, and finally, to discuss about some current directions in QD-LEDs research that concentrate on the realization of the next-generation displays and high-quality lighting with superior color gamut, higher efficiency, and high color rendering index.

Keywords: quantum dots, quantum dot-based light-emitting diodes, display technology, lighting technology

1. Introduction

Quantum dots (QDs) have attracted interest in the fields of optical applications such as quantum computing, biological, and chemical applications.

In contradistinction to the traditional fluorophores, QDs have unique optical and electronic features, which comprise high quantum yields, high molar extinction coefficients, large effective Stokes shifts [1, 2], broad excitation profiles, narrow/symmetric emission spectra, high resistance to reactive oxygen-mediated photobleaching [2, 3], and are against metabolic degradation [4, 5].

QDs are fluorophore nanocrystals whose excitation and emission is basically distinct than classical organic fluorophores [2]. The unique properties of QDs appear almost exclusively due to the size regime in which they exist [6]. QDs obey the quantum mechanical principle of three-dimensional confinement of the charge carriers (electrons, holes) that determine novel quantum phenomena and tunable optical properties, which are sensitive to the size, shape, and material composition of the QDs [7].

QDs have an intrinsic energy bandgap that decides required wavelength of radiation absorption and emission spectra. The bandgap energy increases with the decrease in the dimension of the QD [8]. The color of the light which a QD emits is directly connected to its size; the bigger dots cause longer wavelengths, lower frequencies, and redder light while the smallest dots produce shorter wavelengths, higher frequencies, and bluer light [9–11]. This dimension dependence permits the modulating of the bandgap energy by varying the size of the QD [6, 12].

The wavelength of fluorescence depends on the bandgap, and consequently, it is determined by the dimension of the QD [6, 12]. The ability to adjust and to control the size of the QDs is relevant and advantageous for many new applications.

Due to their highly tunable properties, QDs have an abundance of applications in a diversity of fields. QD could offer a choice for commercial applications such as display technology. QDs are also brighter than a competing technology that is known as organic light-emitting diode (OLED) displays and could eventually make OLED displays outdated [9]. An OLED is fabricated with organic compounds that light up when fed electricity [13].

An important drawback of the OLED technology is the lack of trust and cheap patterning methods for various color pixels [14]. Because OLEDs are composed of small molecule organics, they are not consistent with the classical lithographical patterning techniques that necessitate exposure to solvents which completely deteriorate the structures of OLED [14]. QDs are the reliable solutions for flat-panel TV screens, digital cameras, mobile phones, and personal gaming equipments because QDs could assist large, flexible displays and would not deteriorate as easily as OLEDs [15].

According to the synthetic method, the QDs can be categorized into epitaxial and colloidal QDs. Colloidal QDs are made through chemical synthesis and are composed of a small inorganic semiconductor core, an inorganic semiconductor shell with a broader bandgap, and a coating of organic passivation ligands [16, 17]. The optical properties of the original QDs, which contain a core, can be improved by using the coating of higher bandgap materials or passivation of the exterior part of the core [16, 18]. QD technology is used to filter light from light-emitting diodes (LEDs) to backlit liquid crystal displays (LCDs). With the recent enhancements introduced by the usage of the QDs to backlighting technology, LED/LCD TVs are much better today than they were just few years ago [13].

The use of QDs for the improvement of the LED backlighting leads to the enhancement of the useful light throughput and to the providing of a better color gamut [15]. This QD technology system sends the light from a blue LED through the medium of a tube filled with red and green QDs, so that at the other end of the tube, this blend of blue, green, and red light incurs less absorption of unwished colors by the color filters behind the LCD screen [13, 19]. The

http://dx.doi.org/10.5772/intechopen.69177

first TV manufacturer that presented the achievement of this new type of technology called Triluminos quantum dot display technology was Sony in 2013 [13].

A classic light-emitting diode (LED) is made of some materials that are chosen to emit the required color light and are arranged in layers in a structure named a device stack. The dimension of the total width of this device stack is around 10 μm [21].

2. Evolution of quantum dot-based light-emitting diodes

Due to the multiple advantages of using QDs and their applications in optoelectronic instruments like LEDs, the scientists have created quantum dot-based light-emitting diode (QD-LED) with the improved efficiency and flexibility. QD-LED represents the following generation's display technology after OLED displays [20–22]. QD-LEDs are a form of light-emitting technology for creating large-area displays that could have applications for TVs, cell phones, and digital cameras [15, 20–22].

The structure of QD-LED is analogous to the fundamental design of OLED, with the difference that the light emitting is the QDs, such as cadmium selenide (CdSe) nanocrystals [20, 21].

A classical QD-LED is composed of three layers: one inner layer of QDs as an emissive layer, one outer layer that transports electrons, and one outer layer that transports holes. After applying an electric field on the outside layers, electrons and holes shift in the layer of QD, where they are captured by QD and recombine, emitting photons [22]. Due to the multiple advantages of using the colloidal QDs, the colloidal QDs are a promising way for making QD-LEDs. A great effect of an increased recombination efficiency is obtained by constructing an emissive layer in a single layer of QDs, so that the electrons and holes may be moved directly from the surfaces of electron-transport layer and hole-transport layer [15, 20, 21]. For the definition of the performance of a QD-LED is used the external quantum efficiency (EQE), which is the term that designates the number of photons emitted from the device per electron. For the methodical progress of QD chemistries and active layer designs as well as new device architectures for high-performance QD-LEDs, it is important to discover and to analyze the fundamental causes of inefficiency and to suggest potential solutions [23].

The investigation of the efficiency of the light generation process in the QD-LEDs is an important criterion for achieving high-performance QD-LEDs [24–26].

In 1907, the British scientist named Henry Joseph Round reported light emission from a crystal detector; thus, the idea of light-emitting diode was introduced [27, 28]. In 1927, the Russian researcher Oleg Vladimirovich Losev published a paper about the first light-emitting diode [27, 29].

In 1955, Rubin Braunstein, who worked at Radio Corporation of America, discovered that some common diodes emit infrared light when connected to a current [30, 31]. Also, Rubin Braunstein and Egon Loebner reported in 1958 a green LED which was realized from a lead antimonde/germanium alloy [32]. In 1962, the researcher Nick Holonyak Jr., who is called as "the father of the light-emitting diode," created the first practical visible-spectrum (red) GaAsP LED at General Electric Company in New York [33]. In 1964, IBM introduced the using of LEDs on circuit boards in an early mainframe computer [31].

Thomas P. Pearsall reported the first high-brightness and high efficiency LED in 1976, for utilization with fiber optics in telecommunications [34]. Akasaki et al. [35] made the first blue LED in 1992 based on GaN with efficiency of 1%.

Shuji Nakamura realized the first high-brightness blue LED in 1979 at Nichia Corporation laboratory, but it was too expensive for commercial use until 1994 [36, 37]. Then, Shuji Nakamura was awarded the 2006 Millennium Technology Prize for the development of a white LED [38]. The possibility of color displays with blue, red, and green LEDs was advantageously accomplished [39]. Isamu Akasaki, Hiroshi Amano, and Shuji Nakamura have won the Nobel Prize in Physics in 2014 for the invention of the blue LED [40].

An actual impediment for the development of the performance of the LED is an insufficient understanding of the contribution of some extrinsic elements, such as non-radiative recombination at surface defects versus intrinsic processes, such as multicarrier Auger recombination [41].

In recent years, there have been a lot of research to enhance the quality of LEDs, so recently were introduced QD-LEDs that have the attractive features which correlate the excited state dynamics of structurally engineered QDs with their emissive performance inside LED.

In this regard, some important results of these efforts are the latest demonstrations of QD-LEDs with achievement surpassing that of fluorescent OLEDs that were seriously investigated for a minimum of two decades [26, 41, 42].

QD-LEDs not only reduce the consumption of energy but also show high color purity. Studies reported that QD-LEDs exhibit the ability to be more than twice as power efficient than OLEDs at the same color purity [43, 44]. QD-LEDs have the advantages of foldability and their wide application for next-generation electronic displays and optical communication technology [45].

QD-LEDs exhibit pure and saturated emission colors with narrow bandwidth. In QD-LEDs, the emission color is powerfully directed by the dimension of the used QD due to the confinement effects [44]. It has been proven that QD-LEDs present a 30–40% luminance efficiency advantage above OLEDs for the same color point [43, 44].

QD-LEDs offer several promising features, such as size-dependent emission wavelength, narrow emission spectrum, high efficiency, flexibility, and low-processing cost of organic light-emitting device [43]. The luminaire manufactural cost is diminished due to the capability to imprint large-area QD-LEDs on ultra-thin transparent or flexible substrates [44].

Alivisatos and his colleagues realized the first QD-LED in 1994 [46]. This kind of QD-LED is consisted of a common bilayer structure including an indium tin oxide (ITO) anode, plain CdSe QDs, a *p*-paraphenylene vinylene (PPV) layer, and Mg cathode. Another paper from two research groups at MIT reported a single-layer CdSe-QD-LED with the nanocrystals embedded into an organic polymer matrix [47]. These devices exhibited a very small value of EQEs of 0.001–0.01% [47, 48] and 0.0005% [47, 49].

In superior performance QD-LEDs, the utilized QDs are core-shell type with some structure gradient from core to shell and with various capping ligands or mixed emitting coats [16, 49, 50]. The researchers found different types of QD-LEDs which are divided into four categories of devices based on their design [16, 49, 50]: type I (QD-LEDs with polymer charge transport coats), type II (QD-LEDs with organic small molecule charge transport coats), type III (QD-LEDs which are composed of inorganic charge transport coats), and type IV (QD-LEDs which contain an inorganic metal oxide semiconductor as the electron transport layer-ETL- and an organic semiconductor as the hole transport layer-HTL). In 2002, it was discovered the first type II QD-LED, which was made up of a single monolayer of QDs, was sandwiched between two organic thin films [51].

A drawback to be mentioned to the previously indicated study [51] is the utilization of the organic charge support layers in QD-LEDs that creates an unwanted contribution to the light emission of the LED. This undesired emission perturbs the color purity when a saturated monochromatic emission is wanted. Several authors have shown that in certain cases the perturbing emission could be eliminated [51–53] and in other cases, it could be utilized to build an efficacious white QD-LED [54, 55]. The researchers established that the highest efficiency devices with the best construction architecture are QD-LEDs from type IV class. This type of QD-LED is a hybrid tool which comprises an inorganic metal oxide semiconductor such as the ETL and an organic semiconductor such as the HTL.

The usage of the inorganic coat produces substantial advantages for device stability in air [56, 57] and maximum current density, whereas the use of the organic coat provides a great tunability and an easy processing [16].

In 2007, Anikeeva et al. [58] reported an efficacious spectrally broad electroluminescent QD-LED with spectral emission tunable across the Comission Internationale de l'Eclairage (CIE) color space. More exactly, the authors describe in this paper LEDs with a broad spectral emission which is produced by electroluminescence from a mixed monolayer of red, green, and blue emitting colloidal QDs in a hybrid organic/inorganic QD-LED. Concurrent electroluminescence of numerous color QDs leads to the evolution of tunable LED colors such as white QD-LEDs [58]. The number of QDs colors which can be utilized in a single device is practically boundless; thus, it is possible to obtain higher color rendering and imitate the solar color temperature with the help of QD-LEDs.

Yang and coworkers [59, 60] demonstrated a full range of blue, green, and red quantum dot-based light-emitting devices exhibiting EQEs above 10%. These devices showed low turn-on voltages and saturated pure colors. It has been reported that the values of the lifetimes for the green and red devices are greater than 90,000 and 300,000 h, respectively.

NanoPhotonica is a company that offers advanced and original nanomaterials and manufacturing methods which assure for electronic displays to show a high resolution, pure, bright colors, and an improved efficiency for an important low cost of production. In 2015, NanoPhotonica exhibited enhanced efficiency for blue and green QD-LEDs [60, 61].

The contribution here reported that by using a promising design strategy for QD synthesis and device fabrication methods, a high value of 21% EQE for the green QD-LEDs and a value of 11.2% EQE for the blue QD-LEDs were discovered. The value of 21% EQE in the case of the green QD-LED is the greatest recorded efficiency of any color QD-LED and is the same value as in the case of vacuum-deposited red and green OLEDs utilized in AMOLED display technology which is commercially available [49].

In the following section, the current status of the applications of QD-LEDs and a summary of the issues concerning the limiting of the applicability of QD-LEDs are discussed.

With the aim of achieving a well-designed QD-LED appropriate for general lighting applications, two relevant criteria such as color rendering index (CRI) and the correlated color temperature (CCT) must be discussed in this case. The CCT represents a measure of the light source color appearance described by the vicinity chromaticity coordinates at the blackbody's locus as a one number rather than the two needed to specify a chromaticity [62, 63]. CRI is defined like the measurement of how the colors look under a light source in comparison with sunlight.

QD-LEDs have already exhibited important advantages in general lighting with higher efficiency and better rendering ability [48, 64, 65]. QD-LEDs have enjoyed a lot of attention as promising devices for next-generation displays. Regarding the usage of QD-LEDs in the area of the display technology, one of the most significant parameters for the characterization of the display devices is color gamut [63]. In this field of applications of QD-LEDs the chromaticity diagrams and color gamut standards in order to measure the purity of the color are utilized.

In this regard, it is noteworthy that the International Commission on Illumination (CIE) created the chromaticity diagram named CIE 1931 color space. QD-LED device represents a competitive backlight resolving for the next-generation LCDs because unlike traditional backlight solutions, the QD backlight provides a larger color gamut with a value more than 115% National Television System Committee (NTSC) in CIE 1931 color space and with a value more than 140% NTSC in CIE 1976 [62, 65]. These evaluation criteria listed above for QD-LEDs devices should be considered to create attractive next-generation display applications and superior quality lighting applications with better color gamut, higher efficiency, and high CRI [62, 65].

3. Actual applications of quantum dot-based light-emitting diodes

QD-LEDs have proved impressive outcomes for medical field, lighting, and display applications.

3.1. Quantum dot-based light-emitting diodes for phototherapy

In this part, an important and new application of the use of QD-LEDs in phototherapy is summarized [66]. Phototherapy or light therapy consists of the dermal exposure to light for the treatment of the various medical disorders. Light therapy is used to treat the skin disorders (psoriasis, acne vulgaris, eczema, skin cancer, wound healing), neonatal jaundice, circadian rhythm disorders, and tumors. The researchers [66] have created two medical dressings for phototherapy, each of the two medical dressings include an occlusive layer and translucent layer. For one medical dressing, QD-LEDs chips are designed within the occlusive layer and covered with a translucent layer, so as to furnish a characteristic wavelength of light for utilization in phototherapy. The second device, which was discovered by the same authors [66], consists of an occlusive layer and a translucent layer with QDs that are enclosed in a layer or both layers.

3.2. Quantum dots as electroluminescent light sources

Solid-state lighting (SSL) represents another field in energy and environmental sustainability, in which QDs can have a remarkable price and a great advantage compared to the actual state of the art. Therefore, QDs are considered valuable materials for displays that are suitable for the incorporation into SSL technologies as downconversion phosphors or electroluminescent phosphors [49]. In last years, much research has been achieved on QD-LEDs as the optimal choice for SSL applications [49, 67]. The greatest assets of QD-LEDs as electroluminescent light sources for SSL technologies are their low cost, high efficiency production compatibility, flexible and versatile form factors, and the capacity to be a light source spread over a large area than a point light source. There are some performance products of this kind which become a commercial reality. For example, two UK companies [64] named Cumbria LED lighting company Marl and Manchester-based Nanoco (Nanoco is the global leader in the evolution and the fabrication of the cadmium-free quantum dots-CFQD) performed the world's first CFQD quantum dot LED lighting product, the Orion QD. QDs that are produced by Nanoco are cadmium-free, absorb light in a broad wavelength series, comprising blue, and emit at a color conditioned by their dimension. In this case, light is only converted to where it is needed.

This type of product finds its application in horticulture because the nanoparticle luminaires produce light comprising blue and red but not green that corresponds to the absorption of the chlorophyll, with none of the light it reflects [49, 68]. Another high-quality example is represented by a device from Zylight, namely the multiple award-winning F8 100 LED Fresnel which is the next generation of Fresnel light and encloses QDs into its fixture [49, 69]. More exactly, in compliance with Zylight, the F8 includes a certain mix of QDs with traditional phosphor and exhibits a value of the CR) of up to 97 and a quality of light adjusted only by traditional sunlight and incandescent bulbs [69, 70]. It is expected that with the progress of the research in the field of electroluminescent QD-LEDs, these devices bring substantial improvements for the SSL industry.

The recent studies conducted on the LED base next-generation lighting show remarkable results in the area of lighting for the energy saving for the world warming prevention. In this regard, Song [71] realized a multifunctional LED light source as investigation about the lighting installation using the QD. Due to the fact that artificial light is utilized in different areas such as the increase of the efficiency lighting, the rise of the growth of plant, and the prevention of the disease, LED is a great choice to be extensively used as artificial light in factory plant. Moreover, this LED mentioned before is an energy saving-device and diminishes the emissions of the greenhouse gas. The author presented the achievement [71] of a manufactured LED lamp by the use of QDs as phosphor, in order to be applied like an illumination technology for plant growth used in a multi-wavelength QD-LED device.

In 2013, Pickett et al. [72] realized an invention that has to do with the QD-LEDs useful for plant, algae, and bacterial growth applications. To overcome the disadvantages of using LEDs, the inventors have suggested the utilization of QD-LEDs as the optimal choice. Thus, the researchers presented in their work that QD-LEDs utilized a primary light source which is a solid-state LED, with blue or UV light, and a secondary light source (which downconverts the primary light) that comprises one or more QD components. In their paper, Pickett et al. [72] described that for the optimization of the plant growth in the agricultural and horticultural field, for the improvement of the process of the growth of algae, and for the stimulation of the photosynthetic bacterial growth in bioremediation goals, the QD-LED lighting systems are utilized. In contrast to the solid-state LED lighting, the QD-LED lighting system offers a less costly choice, emits less heat which could harm the plants and other photosynthetic organisms, can offer a greater light intensity, emits light at wavelengths better geared toward the promotion of the growth of bacteria and algae, thereby minimizing energy losses, and exhibits a high-energy efficiency. Thus, the value of the energy efficiency of QD-LEDs used in this case is in the range of 30–70 lm/W, in contrast to 10–18 lm/W for incandescent bulbs and 35–60 lm/W for fluorescent lamps. Another important advantage of the use of the QD-LED lighting systems is that due to the easy wavelength adjustment of QDs, the emission wavelength of the QD-LED can be simply changed to correspond to a diversity of various photosynthetic bacteria. The QD-LED systems described in this invention can be utilized in a wide variety of applications.

There are two kinds of QD-LEDs, and the dissimilates between these two kinds of QD-LEDs are that the first category is based on photo-excited QDs (photoluminescence QD-LEDs) and the second category of QD-LEDs relies on electro-excited QDs (electroluminescence QD-LEDs) [62]. The most frequently utilized type of QD-LEDs in applications is the photoluminescence QD-LEDs.

Klimov and colleagues [41, 73, 74] from the Nanotechnology and Advanced Spectroscopy Team at Los Alamos National Laboratory reported some substantial advances in the domain of the applications of QD-LEDs. Klimov believes that QD-LEDs can probably offer many benefits over traditional lighting technologies, such as incandescent bulbs, particularly in the fields of efficiency, functioning lifetime, and the color quality of the emitted light. Advanced investigations, made by Klimov and his team from Los Alamos, allow that less wasteful fluorescent light sources quickly substitute the incandescent bulbs, which are known for the conversion of only 10% of electricity into light and the loss of 90% of it to heat.

The researchers conducted some spectroscopic studies on the QD-LEDs [41, 73, 74] and have shown that the so-called Auger recombination effect has heavily influenced both LED efficiency and the onset of efficiency roll-off at high currents. The team of researchers established two methods that diminished this issue by using hetero-structured quantum dots.

In the last years, there has been much study in the field of LEDs and photovoltaic solar cells (PV SC). Though OLEDs have the guarantee to overcome the traditional LEDs in performance, OLED materials and manufactural processes of them are not sufficiently advanced to offer this economically [75]. By the evolving a hybrid tool, the efficiency can be increased and the manufactural price can potentially be reduced because this hybrid material system is compatible with inexpensive fabrication procedure such as solution processing and roll-to-roll deposition and with patterning methods, enabling multicolor light sources to be prepared on the same substrate by replacing the emissive colloidal QD coating [14, 75].

For example, McCreary developed a hybrid device by combining QDs with conjugated polymers to create a QD-LED. The motivation why this design was chosen is to be able to inkjet print the entire tool, at least the polymer and QD layers [75]. The researcher proposed a structure of the hybrid device which is of type ITO/PEDOT/CdSe QD/Au and is shown in **Figure 1(A)** [75]. In **Figure 1(B)** the energy bandgap structure for the same tool is presented [75].

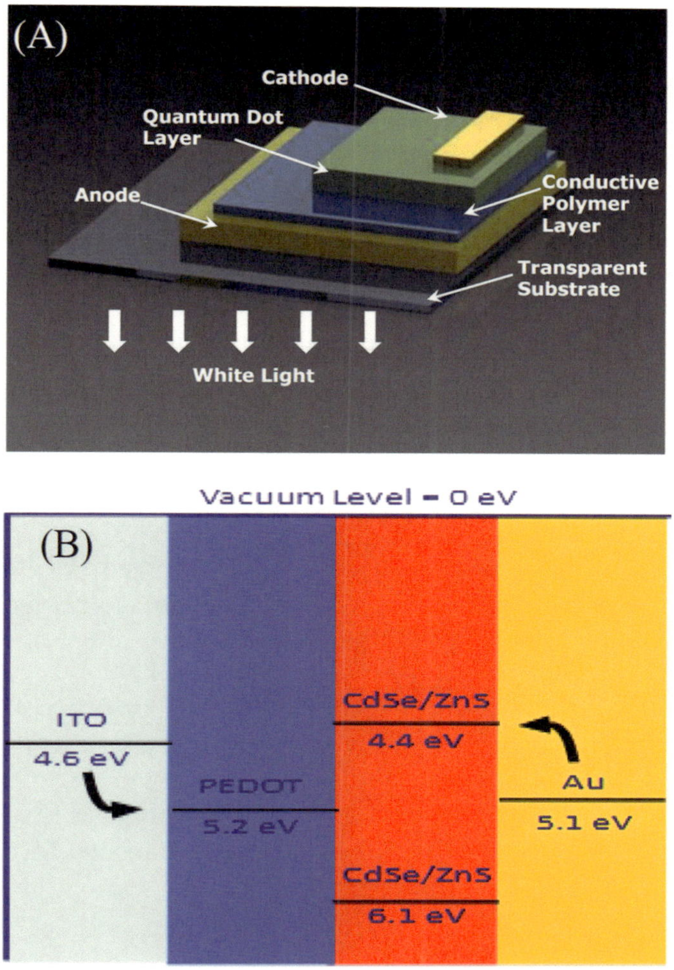

Figure 1. The tool structure of a hybrid LED. (A) Three-dimensional description of the proposed QD-LED [75]. (B) An energy bandgap scheme of the proposed QD-LED and the suggested materials that were used to construct this type of QD-LED [75].

Also, the same author explained the manufacture of a hybrid LED with the structure ITO/PEDOT: PSS/PVK/CdSe QD/Alq3/Al [75]. In order to ease proper hole transport and adequate QD coating, the researcher used a PVK/QD composite solution to make a monolayer layer of QDs using phase separation of the solutes in solution.

This physical modeling of hybrid QD-LEDs of this type, such as those mentioned before [75], makes them applicable to a diversity of hybrid organic QD optoelectronics tools like LEDs, solar cells, photodetectors, and chemical sensors.

3.3. Quantum dot-based light-emitting diodes for near-field scanning optical microscopy

QD-LEDs manufactured on silicon have the potential to be used in nanophotonics, optical micro/nanoelectromechanical systems (MEMS/NEMS), and micrototal analysis systems for real-time biomedical screening [21, 76]. QD-LEDs are of considerable interest for new optoelectronic applications such as that which comprise near-field microscopy beyond the diffraction limit, MEMS-based medical endoscopes for sub-cellular imaging, and compact light-on-chip biosensor and biochips [21].

In his paper, Zhu has investigated the case of a new thin film LED device utilizing nanocrystalline silicon QDs as an emission layer and metal oxide as charge transport layers [77]. Silicon (Si) is notably less costly in comparison with materials like germanium or gallium that are applied for commercial SSL devices and is relatively non-toxic as compared to heavy metal like Cd or Pb. The author developed a thin film LED structure which is based on colloidal silicon nanocrystals using nickel oxide (NiO) and zinc oxide (ZnO) as charge transport layers. The tool that was reported by the researcher is depicted in **Figure 2** [77]. ITO represents the anode, and Aluminum (Al) acts as the cathode. The light is produced when electrons and holes radiatively combine in the silicon nanocrystals (ncSi).

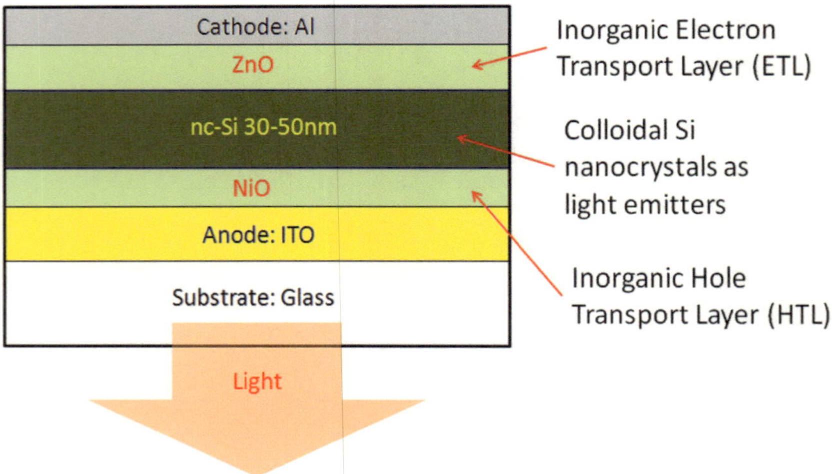

Figure 2. ncSiLED structure with metal oxide charge transport layers [77].

Zhu provided an alternative to organic charge transport layers in OLEDs, demonstrating that metal oxide transport layers based on NiO and ZnO are electrically more conductive than organic charge transport layers regularly found in OLEDs.

Nanocrystalline silicon QDs exhibited tunable luminescent colors. In this chapter, the author showed that nanocrystalline silicon QDs represent a feasible, truly benign, and ecological substitute for the heavy metal (Cd, Pb) QDs, without altering in any way the optical properties [77].

There are a lot of studies that demonstrated a number of procedures of creating QD-LEDs through microcontact printing of QDs on a micromachined silicon probe and resulting in this way a novel generation of highly integrated nano-scale optical fluorescent microscopy [76, 78]. The use of this innovative technology permits to detect the variation of sub-cellular characteristics and to measure the absorption at various wavelengths upon the near-field lighting of individual tumor cells with the aim of the identification of the cancer developmental phase [76].

Another important work in the same area of applications of QD-LEDs relates to the near-field scanning optical microscopy (NSOM) areas of modern investigation with a QD-LED incorporated at the tip of a scanning probe. Hoshino et al. [79] proved near-field fluorescence excitation and imaging using a QD-LED integrated at the tip of a scanning probe. Also, because QD-LEDs have unique properties like well-controlled emission wavelength and narrow bandwidth, they represent a great choice as excitation sources for fluorescence imaging.

In other works [80, 81], the same researchers declared a microcontact printing method which is used to obtain some patterned QD-LEDs on flat silicon substrates. It has been observed that this method is very profitable owing to the connection of the methodology to build silicon-based electronics and MEMS. Hoshino and his team of researchers [79] proved a fluorescence imaging technique, showing that the sensitivity of fluorescence intensity to the QD-LED-QD specimen distance was evaluated down to 50 nm in order. This procedure might be expanded for the unique molecular order measurements [79, 82].

Zhang et al. [83] described a method for the development of a new near-field scanning probe with sub-diffraction-limit resolution by producing a nanometer-sized light source on a patterned probe tip and for the use of the probe in order to detect the molecular signatures of the tumors of the breast. In this project, the authors presented scanning fluorescence imaging with a nano-scale light-emitting diode incorporated at the tip of a silicon microprobe. More accurately, the researchers constructed and studied the QD-LED for bioimaging applications and examined the fluorescence imaging with the on-probe nano-scale QD-LED. It has been shown that QD-LEDs function like near-field excitation sources to enlighten fluorescently labeled cancer cells such as breast cancer cells named MDA435, prostate cancer cells called PC3, and circulating tumor cells in blood. In this chapter [81], a QD-LED at the tip of a micromachined silicon scanning probe was built; the very small size of the QD-LED makes it usable mainly for sub-wavelength optical measurements like in the case of NSOM measurements.

4. Conclusions

The area of QD-LED technology has made immense strides in the past 10 years and the demand for lower price and higher efficiency devices with raised functionality will continue to lead the novelty.

Among all the emerging display and lighting technologies briefly considered in this chapter, QD-LED technologies are by far the most nascent. Like technology, it is a direct challenge to OLEDs.

In this chapter, the recent developments in applications of QD-LEDs and some QD-LEDs qualities in display and lighting applications including their color tunability, durability, and high luminescence efficiency have been reviewed and discussed.

In this work, unique features of QD-LEDs applications for fundamental research and industry which will indubitably spectacularly bring novel design possibilities for the next-generation displays and solid-state lighting in the years to come are presented.

Author details

Anca Armăşelu

Address all correspondence to: anca_armaselu@yahoo.com

Department of Electrical Engineering and Applied Physics, Faculty of Electrical Engineering and Computer Science, Transilvania University of Brasov, Brasov, Romania

References

[1] Mazumder S, Dey R, Mitra MK, Mukherjee S, Das GC. Review: Biofunctionalized quantum dots in biology and medicine. Journal of Nanomaterials. 2009;**2009**:1-17. DOI: 10.1155/2009/815734

[2] Deerinck TJ. The application of fluorescent quantum dots to confocal, multiphoton and electron microscopic imaging. Toxicologic Pathology. 2008;**36**(1):112-116. DOI: 10.1177/0192623307310950

[3] Medintz IL, Uyeda HT, Goldman ER, Mattoussi H. Quantum dot bioconjugates for imaging, labelling and sensing. Nature Materials. 2005;**4**:435-446. DOI: 10.1038/nmat1390

[4] Jaiswal JK, Mattoussi H, Mauro JM, Simon SM. Long-term multiple color imaging of live cells using quantum dot bioconjugates. Nature Biotechnology. 2003;**21**(1):47-51. DOI: 10.1038/nbt767

[5] Ballou B, Lagerholm BC, Ernst LA, Bruchez MP, Waggoner AS. Noninvasive imaging of quantum dots in mice. Bioconjugate Chemistry. 2004;**15**(1):79-86. DOI: 10.1021/bc03453y

[6] Melville J. Optical properties of quantum dots [final paper]. UC Berkeley College of Chemistry; 2015. Available from: https://www.ocf.berkeley.edu/~jmlvll/lab-reports/quantumDots/quantumDots.pdf

[7] Moody G. Confinement effects on the electronic and optical properties of semiconductor quantum dots revelead with two-dimensional coherent spectroscopy [thesis]. University of Colorado; 2013. Available from: https://jila.colorado.edu/sites/default/files/assets/files/publications/Thesis_moodyg.pdf

[8] McDaniel J. Quantum Dots: Science and Applications. Available from: http://ion.chem.usu.edu/~tapaskar/James-Quantum%20Dots%20Seminar.pdf

[9] Woodford C. What are Quantum Dots? [Internet]. 2016. Available from: http://www.explainthatstuff.com/quantum-dots.html

[10] Parak WJ, Pellegrino T, Plank C. Labelling of cells with quantum dots. Nanotechnology. 2005;**16**(2):9-25. DOI: 10.1088/0957-4484/16/2/R01

[11] Kippeny T, Swafford LA, Rosenthal SJ. Semiconductor nanocrystals: A powerful visual aid for introducing the particle in a box. Journal of Chemical Education. 2002;**79**(9):1094-1100. DOI: 10.1021/ed079p1094

[12] Parak WJ, Manna L, Rimmel Ch, Gerion D, Alivisatos P. Quantum dots. In: Schmid G, editor. Nanoparticles—From Theory to Application. Weinheim: John Wiley and Sons; 2004. 447 p. ISBN: 3-527-30507-6

[13] Denison C. OLED vs LED: Which is the Better TV Technology? [Internet]. 2016. Available from: http://www.digitaltrends.com/home-theater/oled-vs-led-which-is-the better-tv-technology/

[14] Anikeeva PO. Physical properties and design of light-emitting devices based on organic materials and nanoparticle [thesis]. Massachusetts Institute of Technology; 2009. Available from: https://nanohub.org/resources/9435/download/Polly_thesis.pdf

[15] Quantum Dot Display [Internet]. 2013. Available from: https://en.wikipedia.org/wiki/Quantum_dot_display#cite_note-New_Scientist-1

[16] Shirasaki Y, Supran GJ, Bawendi MG, Bulović V. Emergence of colloidal quantum-dot light-emitting technologies. Nature Photonics. 2013;**7**:13-23. DOI: 10.1038/nphoton.2012.328

[17] Amini P, Dolatyari M, Rostami G, Rostami A. High througput quantum dot based LEDs. In: Eissa MM, editor. Energy Efficiency Improvements in Smart Grid Components. Rijeka: InTech; 2015. pp. 293-324. DOI: 10.5772/59092

[18] Kim S, Im SH, Kim SW. Performance of light-emitting-diode based on quantum dots. Nanoscale. 2013;**5**:5205-5214. DOI: 10.1039/C3NR00496A

[19] Perry TS. CES 2015: What the Heck are Quantum Dots? [Internet]. 2015. Available from: http://spectrum.ieee.org/tech-talk/consumer-electronics/audiovideo/what-the-heck-are-quantum-dots

[20] NanoPhotonica [Internet]. 2015. Available from: http://nanophotonica.net/technology/toggle-id-2

[21] Hussain Z. Introduction to Nanobiotechnology [Internet]. 2015. Available from: http://www.slideshare.net/zohaibkhan404/quantum-dot-light-emitting-diode

[22] Zyga L. Quantum Dot LED Approaches Theoretical Maximum Efficiency [Internet]. 2013. Available from: https://phys.org/news/2013-05-quantum-dot-approaches-theoretical-maximum.html

[23] Bozyigit D, Wood V. Challenges and solutions for high-efficiency quantum dot-based LEDs. MRS Bulletin. 2013;**38**:731-736. DOI: 10.1557/mrs.2013.180

[24] Anikeeva PO, Madigan CF, Halpert JE, Bawendi MG, Bulović V. Electronic and excitonic processes in light-emitting devices based on organic materials and colloidal devices based on organic materials and colloidal quantum dots. Physical Review B. 2008;**78**(8):085434-085441. DOI: 10.1103/PhysRevB.78.085434

[25] Wood V, Panzer MJ, Halpert JE, Caruge J-M, Bawendi MG, Bulović V. Selection of metal oxide charge transport layers for colloidal quantum dot LEDs. ACS Nano. 2009;**3**(11):3581-3586. DOI:10.1021/nn901074r

[26] Mashford BS, Stevenson M, Popovic Z, Hamilton C, Zhou Z, Breen C, Steckel J, Bulović V, Bawendi M, Coe-Sullivan S, Kazlas PT. High-efficiency quantum-dot light emitting devices with enhance charge injection. Nature Photonics. 2013;**7**:407-412. DOI: 10.1038/nphoton.2013.70

[27] Wac Lighting. Responsible Lighting. LED/OLED: Technical Training and Applications [Internet]. 2013. Available from: http://iie.ciapr.org/actividades/seminarios/2009/Presentacion LED-OLED.pdf

[28] Round HJ. A note on carborundum. Electrical World. 1907;**19**:309-310

[29] Losev OV. Luminous carborundum (silicon carbide) detector and detection with crystals. Telegrafiya i Telefoniya bez Provodov. 1927;**44**:485-494

[30] Braunstein R. Radiative transitions in semiconductors. Physical Review. 1955;**99**:1 892-1893

[31] Paisnik K, Rang G, Rang T. Life-time characterization of LEDs. Estonian Journal of Engineering. 2011;**17**(3):241-251. DOI: 10.3176/eng.2011.3.05

[32] Braunstein R, Loebner EE. Semiconductor device for generating modulated radiation. RCA Corp, assignee, U.S. Patent 3102201 [Issued: 27 August 1963]

[33] Holonyak N Jr, Bevaqua SF. Coherent (visible) light emission from $Ga(As_{1-x}P_x)$ junctions. Applied Physics Letters. 1962;**1**(4):82-83. DOI: 10.1063/1.1753706

[34] Pearsall TP, Miller BI, Capik KJ, Bachmann KJ. Efficient lattice-matched double-heterostructure LED's at 1.1. μm from $Ga_xIn_{1-x}As_yP_{1-y}$. Applied Physics Letters. 1976;**28**(9):499-501. DOI: 10.1063/1.88831

[35] Akasaki I, Amano H, Itoh K, Koide N, Manabe K. GaN-based ultraviolet/blue light emitting devices. Inst. Phys. Conf. Ser. 1992;**129**:851-856

[36] Nakamura S, Mukai T, Senoh M. Candela—class high—brightness InGaN/AlGaN double—heterostructure blue-light-emitting diodes. Applied Physics Letters. 1994;**64**: 1687-1689

[37] History of Lighting [Internet]. Available from: http://www.historyoflighting.net/light-bulb-history/history-of-led/

[38] Shuji Nakamura's Millennium Prize a Brilliant Tribute. In: LED Lighting Luminaires and Fixtures. 2006. Available from: http://www.solidstatelightingdesign.com/shuji-nakamuras-millenium-prize-a-brilliant-tribute/

[39] Muthu S, Schuurmans FPJ, Pashley MD. Red, green, and blue LEDs for white light illumination. IEEE Journal of Quantum Electronics. 2002;**8**:333-338

[40] Webb J. Invention of Blue LED Wins Physics Nobel [Internet]. 2014. Available from: http://www.bbc.com/news/science-environment-29518521

[41] Bae WK, Park Y-S, Lim J, Lee D, Padilha L, McDaniel H, Robel I, Lee C, Pietryga JM, Klimov VI. Controlling the influence of Auger recombination on the performance of quantum-dot light-emitting diodes. Nature Communications. 2013;**4**(2661):1-8. DOI: 10.1038/ncomms3661

[42] Kwak J, Bae WK, Lee D, Park I, Lim J, Park M, Cho H, Woo H, Yoon DY, Char K, Lee S, Lee C. Bright and efficient full-color colloidal quantum dot light-emitting diodes using an inverted device structure. Nano Letters. 2012;**12**(5):2362-2366. DOI: 10.1021/NL3003254

[43] Altavilla C, Ciliberto E, editors. Inorganic Nanoparticles: Synthesis, Applications, and Perspectives. Boca Raton, FL: Taylor & Francis; 2011. 547 p. ISBN: 9781439817612-CAT#K10915

[44] Infos and News about QLEDs Quantum Dot Light Emitting Diodes [Internet]. 2013. Available from: http://www.qled-info.com/introduction/

[45] Yang XY, Mutlugur E, Zhao YB, Gao Y, Leck KS, Ma YY, Ke L, Tan ST, Demir HV, Sun XW. Solution processed tungsten oxide interfacial layer for efficient hole-injection in quantum dot light-emitting diodes. Small. 2014;**10**(2):247-252. DOI:10.1002/smll.201301199

[46] Colvin VL, Schlamp MC, Alivisatos AP. Light-emitting diodes made from cadmium selenide nanocrystals and a semiconducting polymer. Nature. 1994;**370**(6488):354-357. DOI: 10.1038/370354a0

[47] Dabboussi BO, Bawendi MG, Onitsuka O, Rubner MF. Electroluminescence from CdSe quantum-dot/polymer composites. Applied Physics Letters. 1995;**66**(11):1316-1318. DOI: 10.1063/1.113227

[48] Erdem T, Nizamoglu S, Sun XW, Demir HV. A photometric investigation of ultra-efficient LEDs with high color rendering index and high luminous efficacy employing

nanocrystal quantum dot luminophores. Optics Express. 2010;**18**(1):340-347. DOI: 10.1364/OE.18000340

[49] Kitai A, editor. Materials for Solid State Lighting and Displays. Chichester: Wiley; 2017. 384 p. ISBN: 978-1-119-14058-0

[50] Supran GJ. QLEDs for display and solid-state lighting. MRS Bulletin. 2013;**3819**:703-711. DOI: 10.1557/mrs.2013.181

[51] Coe S, Woo W-K, Bawendi M, Bulović V. Electroluminescence from single monolayers of nanocrystals in molecular organic devices. Nature. 2002;**420**(6917):8100-8803. DOI: 10.1038/nature01217

[52] Tan Z, Zhang F, Zhu T, Xu T. Bright and color-saturated emission from blue light-emitting diodes based on solution-processed colloidal nanocrystal quantum dots. Nano Letters. 2007;**7**(12):3803-3807. DOI: 10.1021/nl07230s

[53] Zhu T, Shanmugasundaram K, Price SC, Ruzyllo J, Zhang F, Xu I, Mohney SE, Zhang Q, Wang AY. Mist fabrication of light emitting diodes with colloidal nanocrystal quantum dots. Applied Physics Letters. 2008;**92**(2):023111-023113. DOI: 10.1063/1.2834734

[54] Li Y, Rizzo A, Mazzeo M, Carbone L, Manna L, Cingolani R, Gigli G. White organic light-emitting devices with CdSe/ZnS quantum dots as a red emitter. Journal of Applied Physics. 2005;**97**(11):113501. DOI: 10.1063/1.1921341

[55] Chen J, Zhao D, Li C, Xu F, Lei W, Sun L, Nathan A, Sun XW. All solution processed stable white quantum dot light-emitting diodes with hybrid ZnO@TiO_2 as blue emitters. Scientific Reports. 2014;**4**(4085):1-6. DOI: 10.1038/srep04085

[56] Klein M. Quantum Dots for LED Displays [Internet]. 2014. Available from: http://www.materialsforenergytypepad.com/materials/2014/02/quantum-dots-for-led-displays.html

[57] Kumar B, Campbell SA, Ruden PP. Modeling charge transport in quantum dot light emitting devices with NiO and ZnO transport layers and Si quantum dots. Journal of Applied Physics. 2013;**114**(044507):1-6. DOI: 10.1063/1.4816680

[58] Anikeeva PO, Halpert JE, Bawendi MG, Bulović V. Electroluminescence from a mixed red-green-blue colloidal quantum dot monolayer. Nano Letters. 2007;**7**(8):2196-2200

[59] Yang Y, Zheng Y, Cao W, Titov A, Hyvonen J, Manders JR, Xue J, Holloway PH, Qian L. High-efficiency light-emitting devices based on quantum dots with tailored nanostructures. Nature Photonics. 2015;**9**(4):259-266. DOI: 10.1038/nphoton.2015.36

[60] NanoPhotonica [Internet]. 2015. Available from: http://www.nanophotonica.net/about/

[61] Qin W, Yang Z, Jiang Y, Lam JWY, Liang G, Kwok HS, Tang BZ. Construction of efficient deep blue aggregation—induced emission luminogen from triphenylethene for nondoped organic light-emitting diodes. Chemistry of Materials. 2015;**27**(11):3892-3901. DOI: 10.1021/acs.chemmater.5b00568

[62] Xie B, Hu R, Luo X. Quantum dots-converted light-emitting diodes packaging for lighting and display: Status and perspectives. Journal of Electronic Packaging. 2016; **138**(2):020803-020803-13. DOI: 10.1115/1.4033143

[63] Lighting Research Center [Internet]. 2004. Available from: http://www.lrc.rpi.edu/programs/nlpip/lightinganswers/lightsources/whatisCCT.asp

[64] Zhong P, He GX, Zhang MH. Optimal spectra of white light-emitting diodes using quantum dot nanophosphors. Optics Express. 2012;**20**(8):9122-9134. DOI: 10.1364/OE.20.009122

[65] Luo Z, Chen Y, Wu ST. Wide color gamut LCD with a quantum dot backlight. Optics Express. 2013;**21**(22):26269-26284. DOI: 10.1364/OE.21.026269

[66] Harris J, Glarvey P, Masala O, Pickett N, Gresty N. Quantum dot light-emitting diodes for phototherapy. Nanoco Technologies, Ltd.; 2014; WO 2014177943 A2. Available from: http://www.patentsencyclopedia.com/app/20140277297

[67] Supran GJS. Enhancing quantum-dot luminescence in visible and infrared light emitting devices [thesis]. Massachusetts Institute of Technology; 2016. Available from: https://dspace.mit.edu/handle/1721.1/104112

[68] World-First Nano-Particle Luminaires Showcase Growing Benefits [Internet]. 2015. Available from: http//www.en.ofweek.com/new-products/Worl-first-nanoparticle-luminaires-showcase-growing-benefits-36576

[69] The Fresnel Light. Reimagined [Internet]. 2017. Available from: http://www.zylight.com/f8-fresnel/

[70] Allard M. Built Like a Tank: The Zylight F8 High Power LED Fresnel Reviewed [Internet]. 2015. Available from: http://www.newsshooter.com/2015/05/01/built -like-a-tank-the-zylight-f8-led-fresnel/

[71] Song J-W. Grow light for plant factory using quantum dot LED. Journal of International Council on Electrical Engineering. 2016;**6**(1):13-16. DOI: 10.1080/22348972.2016.1138604

[72] Pickett N, Nasaani I, Harris J, Gresty N. Quantum dot led's to enhance growth in photosynthetic organism. Nanoco Technologies, Ltd.; 2013; WO 2013150388 A2

[73] Talapin DV, Steckel J. Quantum dot light-emitting devices. MRS Bulletin. 2013;**38**(9):685-691. DOI: 10.1557/mrs.2013.204

[74] Nanoscale Engineering Boosts Performance of Quantum Dot Light Emitting Diodes [Internet]. 2013. Available from: http://esciencenews.com/articles/2013/10/25/nanoscale.engineering.boosts.performance.quantum.dot.light.emitting.diodes

[75] McCreary MZ. Synthesis of cadmium selenide quantum dots for fabrication of hybrid light emitting diodes [thesis]. University of Louisville; 2014. Available from: http://ir.library.louisville.edu/cgi/viewcontent.cgi?article=1940&context=etd

[76] Gopal A, Hoshino K, Zhang JXJ. Quantum dots light emitting devices on MEMS: Microcontact printing, near-field imaging, and early cancer detection. In: Proceedings of SPIE, International Symposium on Photoelectronic Detection and Imaging 2011: Sensor and Micromachined Optical Device Technologies; Vol. 8191; 2011. p. 819106. DOI: 10.1117/12.901089. Available from: http://proceedings.spiedigitallibrary.org/proceeding.aspx?articleid=1270250

[77] Zhu J. Nanocrystalline silicon quantum dot light emitting diodes using metal oxide charge transport layers [thesis]. University of Toronto; 2012. Available from: https://tspace.library.utoronto.ca/bitstream/1807/42432/6/Zhu_Jiayuan_201211_MASc_thesis.pdf

[78] Hoshino K, Gopal A, Zhang X. Contact printing of quantum dot light emitting diode on silicon probe tip. CLEO/QELS: 2010 Laser Science to Photonic Applications; IEEE; 2010. INSPEC Accession Number: 11428025. DOI: 10.1364/CLEO.2010.CTuNN4. Available from: http://ieeexplore.ieee.org/document/5500224/

[79] Hoshino K, Gopal A, Glaz MS, Vanden Bout DA, Zhang X. Nanoscale fluorescence imaging with quantum dot near-field electroluminescence. Applied Physics Letters. 2012;**101**(4):043118-043118-5. DOI: 10.1063/1.4739235

[80] Gopal A, Hoshino K, Kim S, Zhang X. Multi-color colloidal quantum dot based light emitting diodes micropatterned on silicon. Nanotechnology. 2009;**20**(23):235201-235201-9. DOI: 10.1088/0957-4484/20/23/235201

[81] Gopal A, Hoshino K, Zhan X. Photolithographic patterning of subwavelength top emitting colloidal quantum dot based inorganic light emitting diodes on silicon. Applied Physics Letters. 2010;**96**(13):131109-131109-3. DOI: 10.1063/1.3373832

[82] Hoshino K, Turner TC, Kim S, Gopal A, Zhang X. Single molecular stamping of sub-10-nm colloidal quantum dot array. Langmuir. 2008;**24**(23):13804-13808. DOI: 10.1021/la802936h

[83] Zhang XJ, Ferrari M, Cheng M-C. EPDT: Nano-scale light emitting diode on silicon cantilever for near-field microscopy of nanovectors biodistribution in tissues and living cells. Austin, TX, United States: University of Texas. 2011. Available from: http://grantome.com/ghrant/NSF/ECCS-0725886

Chapter 3

Quantum Dot-Based Light Emitting Diodes (QDLEDs): New Progress

Neda Heydari, Seyed Mohammad Bagher Ghorashi,
Wooje Han and Hyung-Ho Park

Additional information is available at the end of the chapter

http://dx.doi.org/10.5772/intechopen.69014

Abstract

In recent years, the display industry has progressed rapidly. One of the most important developments is the ability to build flexible, transparent and very thin displays by organic light emitting diode (OLED). Researchers working on this field try to improve this area more and more. It is shown that quantum dot (QD) can be helpful in this approach. In this chapter, writers try to consider all the studies performed in recent years about quantum dot-based light emitting diodes (QDLEDs) and conclude how this nanoparticle can improve performance of QDLEDs. In fact, the existence of quantum dots in QDLEDs can cause an excellent improvement in their efficiency and lifetime resulted from using improved active layer by colloidal nanocrystals. Finally, the recent progresses on the quantum dot-based light emitting diodes are reviewed in this chapter, and an important outlook into challenges ahead is prepared.

Keywords: quantum dot, organic light emitting diode, efficiency, lifetime, active layer

1. Introduction

Due to increased population and consumption of more energy, the people of Earth are faced with a serious shortage of energy resources. Therefore, the primary concern of researchers and manufacturers is closely linked to energy consumption. In recent years, a lot of researches are conducted to achieve efficient and low-energy light sources. Inorganic light emitting diode (LED) and organic light emitting diode (OLED) have been introduced as a result of these efforts to achieve solid-state light sources [1–6]. The outdoor application is one of the important markets for LED lighting. For year 2015, the assessment of the total outdoor lighting market was $6.5 billion USD with LEDs. The outdoor lighting market is expected to grow with growth rate about 4% from 2015 to 2021 [7]. LEDs were used in many applications such as television backlight units and illuminated signs. The US Department of Energy has reported that the achievements to the expected developments in LED technology would save 300 TW per hour of electricity [8]. It means that a remarkable strategy needs to be developed for the simple design and better material to reduce the cost of fabrication. According to Stephanie Pruitt report, the packaged LED profits hit 15.4 billion dollars in 2014 and will grow to 22.1 billion dollars in 2019 [9].

2. Why OLEDs?

In recent years, the display industry and lighting panels have been changed. Many researchers are interested in using polymers and organic molecules as emissive layers in these devices to improve their characteristics. One of the most important developments related to OLED technology provides the ability to build flexible (can be deposited onto substrate like plastic), transparent and very thin displays and components. Simply an organic light emitting diode is constructed with a thin film of organic (carbon-based) put between a conductive cathode (electron injection site) and a conductive anode (electron removal site) considering that at least one of the electrodes should be transparent. This thin film is called emitter, which is electroluminescent; it emits light when excited by an electrical current. These organic matters have conductivity levels between insulating and conductive; therefore, they are considered as organic semiconductors. The highest occupied molecular orbital (HOMO) and the lowest unoccupied molecular orbital (LUMO) in organic semiconductors are similar to the conduction and valence bands in inorganic semiconductors. The performance of OLEDs reaches important goals in display technology. In addition, OLEDs have many advantages over both LCDs and LEDs such as thinner, lighter, more flexible and brighter substrate. Moreover, OLEDs unlike the LCDs do not need a backlight and filters; thus, they are very thin, and their construction will be easier and reliable. Due to low energy consumption, OLEDs will be an important advantage for cellphones, which are battery-operated devices. Another important feature of the OLED is the solution-based emitting materials used in their structure. It can be possible to fabricate them into large area by a spin-coating method, which is low-cost fabrication techniques [10]. Also, OLEDs can be produced into large, thin sheet which makes them an interesting choice for industry. In addition, changing information in this technology is in real time, which is faster than LCDs.

The ability of a light source to reveal the colours of objects compared to a natural light source is called colour rendering index (CRI). This parameter is the most important advantage of an OLED in comparison with LED. Consequently, OLEDs have attracted a lot of attention due to light weight and high image quality. These features lead to a wide range of applications in industry, particularly in manufacturing flexible screens and full-colour light emitting pages. But there are still some problems like sensitivity to water vapour. Also, the production costs should be reduced more. Technology of OLED has much room for continuous progress in future. On the other hand, the fabrication process of small-molecule OLEDs is too expensive because thermal deposition with high vacuum is required. However, polymer-based OLEDs (POLEDs) are the good substitute due to their solution process, which makes them more cost-effective. In fact, straightforward way of fabrication is the necessary factor for the low-cost electronic devices. Another advantage of POLEDs is their lower power consumption in comparison with traditional option. Therefore, many researchers are interested in developing POLED technology. Bottom-emitting conventional, bottom-emitting inverted, top-emitting conventional and top-emitting inverted are the four different architectures of POLEDs. Bottom-emitting inverted and top-emitting inverted can increase operational lifetime and reduce the fabrication and operating cost of the device. In addition, top-emitting conventional and top-emitting inverted can increase light out-coupling efficiency.

3. A brief review of OLED development

The first OLED was manufactured in 1987 by Tong in Kodak company [11]. He realized when an electric current is applied to the molecules of the organic material, this material emits green light. This was the first idea about OLEDs. In the first OLED, the structure was built by an indium tin oxide (ITO)/aromatic diamine/8-hydroxyquinoline aluminium (Alq3)/Mg-Al metal electrode. Up to now, the most organic components used in OLEDs are poly(para-phenylenevinylene) (PPV) [12], polyvinylcarbazole (PVK) [13] and aluminium-tris-(8-hydroxyquinoline) (Alq3) [14, 15]. To commercialize the OLEDs, several aspects must be improved. Therefore, during two decades, a lot of efforts have been made to achieve high performance of OLED devices. For example, the ability of charge injection, charge transport and emission of different layers of OLEDs are three important factors in their performance. To improve these factors, much effort has been devoted by researchers. They have tried to find better anode and cathode materials. They have also attempted to synthesize new materials' high emissivity. Therefore, development in synthesis process and application of electron transport materials, modification of surface in hole injection layer and electron injection layer, using high mobility materials in hole and electron transport layer (ETL), doping the high efficiency emitter dopants in emission layers and reducing the barrier to charge carrier injection by increasing the doping level of materials, was received [16, 17].

To achieve high-brightness display, high electron mobility is necessary in electron transport materials. For enhancement of charge injection, scientists try to use different cathode, and simultaneously, they have tried different surface treatments of ITO [18]. It is well known that employing electroluminescence material with high mobility is required for low-power consumption. On the other hand, the voltage can be decreased by doping, but rapid dopant diffusion can create the quenching centres in the emissive layers, which result in reduction of efficiency. Balancing of electron and holes will increase the efficiency of device that can be achieved by controlling the mobility of the transport layers. Therefore, an increase in the exciton recombination probability and control of the carrier accumulation needs to be adjusted for improving the current and power efficiencies by aligning the bands at the interface between the emitting layer (EML) and ETL [19]. Water/oxygen permeability is another that factor must be noticed. Moreover, encapsulation with a barium oxide (BaO) or calcium oxide (CaO) is used in OLEDs, and an acceptable level of water/oxygen permeability is achieved [20]. As mentioned above, the architecture of OLEDs is one of the parameters that need to improve the performance of organic light emitting diode. So far, different structures of the OLEDs are investigated. Scientists have tried to improve the performance and stability of these devices by substituting of alternative material in different layers of OLEDs. For example, carbon nanotubes [21, 22], graphene [23, 24], metal nanomeshes [25, 26], thin metal films [27, 28] and metal nanowires [29, 30] are employed instead of ITO up to now. In addition, Burns et al. have investigated the effect of thermal annealing super yellow emissive layer on efficiency of OLEDs [31]. By annealing of the emissive layer at 50°C, the external quantum efficiency (EQE) of this device reached a maximum of 4.09%.

4. Looking to the future: outlook

OLEDs have been commercialized in tablet, smart watches and smart phones up to now, and they are stable devices with good efficiency. But they still need to achieve more improvements. Higher efficiency, better stability and being more environmentally friendly are some important factors that researchers are trying to improve them. Samsung has manufactured its mobile displays by red, green and blue OLED subpixels, and LG used white emitting OLED material with WRGB colour filters for its TVs. OLED display includes red, green and blue pixels. The most critical issue is the blue gap in OLED materials. Nowadays, display industries use fluorescent materials for blue colour, but the use of fluorescent materials involves with an increase in power consumption. Therefore, new approaches should be introduced in the technology of the OLED display. From technological point of view, the fluorescence, phosphorescence and thermally activated delayed fluorescence (TADF) are three mechanisms to harvest excitons in OLEDs and considered for improvement of their performance. High-performance and low-cost OLEDs are available after discovery of metal-free organic emitters with thermally activated delayed fluorescence (TADF). There are two kinds of TADF emitters named organic and metal-organic. The maximum external quantum efficiency of these OLEDs has been reached to 25% up to now [32]. The efficiency of TADF OLEDs is comparable with phosphorescent OLEDs. The most value of TADF OLED lifetime is reported over 10,000 h. Carbazole [33, 34] or arylamine-type donors [35, 36] are the main organic TADF emitters that are reported up to date. The excited-state lifetime or emission decay time of materials is the important problem that should be solved to commercialize TADF OLEDs. It needs more developments in this field.

5. Structure of OLEDs

Structure of OLED is another factor that can be employed to improve its characteristics. At the device level, each OLED pixel is a p-n junction that emits lights. Top emitting and bottom emitting are two main configurations of the OLEDs. Up to now, top emitting is a common structure that has been used to increase efficiency and the light output of the device by the display industry. When the current source creates potential difference in OLED circuit, a variable voltage between 2 and 10 V is applied between the cathode and the anode, and the flow of electrons from the cathode to the anode is established. After a while the negative charge density in the electron transport layer and the positive charge density in the hole transport layer will increase. Today, some thin layers are used between two conductive layers in order to achieve better performance of OLEDs. The half-life of the device can be reduced by high voltage. Therefore, highly conductive transmission layer is used to reduce injection barriers and achieve low-voltage operation in modern OLEDs. The basic and typical structure of an organic light emitting diode has been shown in **Figure 1**.

The anode is positive compared to the cathode, so the electrons flow from the cathode to the anode. The electrons injected into the cathode are placed in the LUMO level of the organic layer, and they also will withdraw in the HOMO level of the organic layer. Holes arrive from the hole transport layer to the HOMO level. The energy level of the emissive layer should be less than the hole transport layer in order that the injection of the electrons from electron transport layer to LUMO level of the emissive layer to be possible. To penetration the holes into emissive layer, this layer also should have higher HOMO level than HOMO level of the hole transport layer.

Electrostatic forces bring the electrons and holes towards each other and form the excitons in a singlet/triplet ratio of 1:3. It happens near the emissive layer. **Figure 2** shows the population of emitter states by energy transfer from singlet and triplet excitons. It is important to note that the holes are more mobile than electrons in organic semiconductors and arrive to electron transport layer faster. The destruction of this excited state led to radiation in the visible region. If the active layer is phosphorescent, non-radiative triplet excitons may be emitted. The frequency of this radiation depends on the difference between HOMO and LUMO levels of these materials. Because the holes must be logged in the HOMO level of the organic material in emissive layer with energy levels about 5–6 eV, anode with high work function is required till holes will be able to effectively enter to the organic material. Also anode should be transparent in order for the produced photons to be visible. ITO is often used as anode material because it is transparent compared to visible light. In addition, the injection of holes into the HOMO level of the organic layer will be possible due to its high work function.

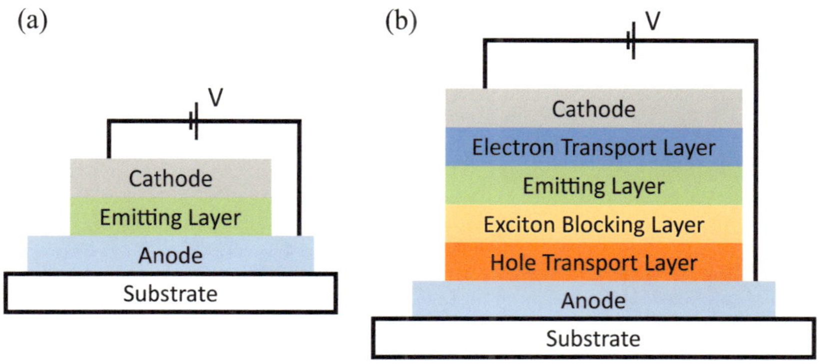

Figure 1. The structure of (a) basic and (b) typical organic light emitting diode.

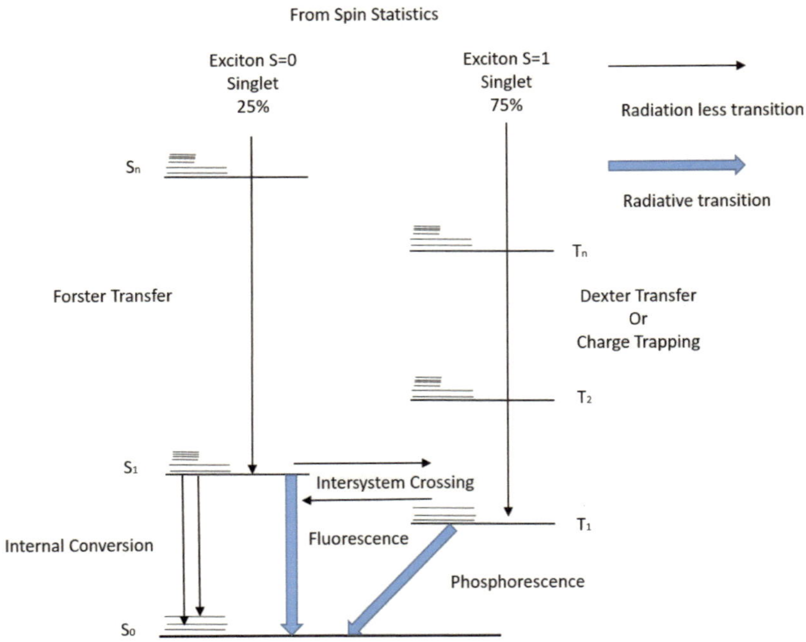

Figure 2. Population of emitter states by energy transfer from singlet and triplet excitons.

To inject electrons into the LUMO level of the organic layer, the work function of the cathode should be low. Calcium and magnesium are two metals used as cathode due to their low work function. But the drawback is that these metals are sensitive to moisture and therefore will reduce the lifetime of the device. To solve this problem, aluminium or various alloys such as Mg/Ag as a cathode are used as cathode [37–39]. **Figure 3** shows the evolution of OLED device structure. The electron and hole transport layers could help to high-speed movement of electrons and holes to meet each other in the emission layer. Electron transport layer prevents the penetration of hole into cathode, and in contrast hole transport layer prevents the penetration of electrons to the anode. The electrons and holes recombine with each other in

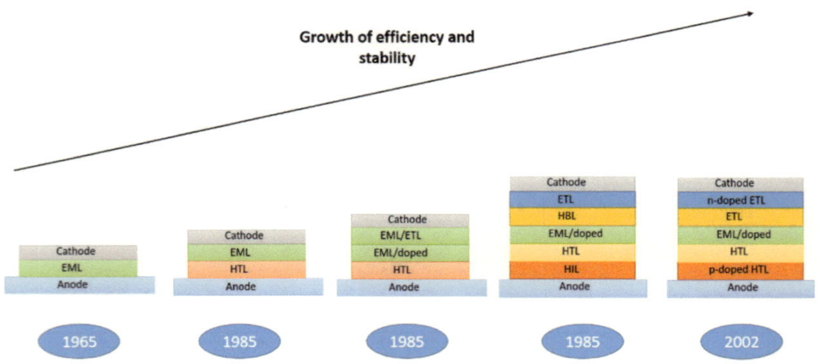

Figure 3. Evolution of OLED device structure.

the middle of emissive layer. The material of the hole transport and the electron transport layers (electron and hole blocking layers) depends on the characteristics of the charge and the values of the HOMO and LUMO levels. Usually PEDOT:PSS is used as a conductive layer in which its HOMO level is between the work function of the ITO and HOMO level of the commonly used polymers in order to reduce the energy barrier of the injecting holes.

6. Different types of the OLEDs

To date, various types of the organic light emitting diodes are presented. The six main types of the OLEDs are passive-matrix OLED (PMOLED), active-matrix OLED (AMOLED), transparent OLED, top-emitting OLED, foldable OLED and white OLED. Each of these types has different kinds of use. PMOLED consists of cathode, organic layers and anode. The manufacture of this type of OLED is easy. They have high-power consumption; therefore, they are effective for small screens. Passive-matrix addressed displays are attractive, as the device construction is relatively simple. AMOLED is composed of cathode, organic molecules and anode layers. The anode layer is established on a thin film transistor (TFT) and formed a matrix. Because of the use of less power, they are suitable for large-sized displays. Substrate, cathode and anode are transparent in the transparent OLED. Top emitting OLED consists of an opacity substrate, and it is suitable for active-matrix design. Foldable OLED has a flexible substrate made of metal or plastic that is very lightweight and durable and can be used in clothes with OLED display. White OLED emits white light that is brighter, more uniform and more efficient than fluorescent lights.

7. Important features of OLEDs

Quantum yields and lifetime are the two important characteristics of an OLED that researchers are trying to improve by different structures and techniques all around the world. Optimizing the balanced charge is an important issue for the lifetime of the device. Also choosing the right ingredients in the manufacture of layers can be useful in improving the performance of the device. OLEDs have the limited peak emission so that the highest peak luminance of OLEDs is at most 500–600 nits. But this value for LCD TVs is about 1800 nits. The amount of the light emitted divided by the amount of the injected current into the piece is called quantum yields of an OLED. High yield achievement and suitable coordinates of colour for display applications are the other important features of an OLED. External quantum efficiency (EQE) can be explained in following formula:

$$\text{External quantum efficiency } (\eta_{EQE}) = \gamma\chi \cdot \eta_{PL}\,\eta_{OC} \quad (1)$$

γ = recombination efficiency of holes and electrons; χ = fraction of excitons with spin allowed optical transitions, created in emissive layer; η_{PL} = photoluminescent efficiency of the emitter; η_{OC} = fraction of emitted photons that are coupled out of the device ($1/2n^2$); n = refractive index of the substrate (glass).

Electrons and holes which meet each other in active layer create different states; about 25% of the excitons are in the singlet states, and the rest of them are in the triplet states. Therefore, the maximum internal efficiency of the OLEDs based on fluorescent molecules is around 25%. It can be possible to improve the efficiency of OLEDs by enhancing spin-orbit coupling and enable emission from the formally forbidden triplet state with the use of phosphors. Totally, display devices are typically assessed through a number of characterization measurements that include colour coordinates (perceived colour), current density (A/cm) versus voltage, luminance (a measure of brightness in cd/m^2) versus voltage, current efficiency (cd/A) versus luminance, power efficiency (lm/W) versus luminance and lifetime (a measure of the stability of the device). Kim and his colleagues in 2014 have shown that approximately 35.6% EQE can be reached by using iridium compounds (HICs). This efficiency is one of the highest external quantum efficiencies achieved to date in the red OLEDs [40]. OLED has already been commercialized; LG company has commercialized the OLED-based TVs. OLEDs are emissive displays, which means they create their own light at each pixel, like CRTs and PDPs. But as mentioned, the possibility of degradation in the presence of moisture and oxygen is the big problem needed to be considered. There are several damaging processes in these materials such as thermal instability, optical and chemical oxidation of the active layer and penetration of the metal from electrodes [41]. So it should be a process for encapsulating structure to protect it from the influence of moisture and oxygen. Also, by replacing the organic materials to inorganic structures, there will be the possibility of a better stability. The best results have been achieved up to date related on the usage of quantum dots based on cadmium. Employing nanoparticles such as oxides and semiconductors to form composite materials is one of the available solutions to improve the stability of these devices [42]. Not only adding nanoparticle can increase the stability of the film, but also it will be possible to control the optical properties by adjusting their size. Thus this would be an appropriate way for optoelectronic applications. For example, the gap between energy levels (luminescence colour) of semiconductor increases by reducing the particle size.

8. A brief review of quantum dot-based light emitting diodes (QDLEDs)

Being cost-effective, much more brightness and more efficient as well as more stable devices made of environmentally sustainable materials are the most important factors that led to the development of the lighting industry. A new candidate for improvement of display industry is emissive layer based on quantum dots (QDs). Quantum dot technology is a novel innovation to help this industry. This technology has also applications in many other markets such as solar cells, biomedical, instrumentation, quantum computers and more. Quantum dot technology seems to offer the biggest colour gamut of the various approaches today. Quantum dots have three key elements to their structure. Core, shell and ligand are the three main properties of the structure of the QDs. The core adsorbs and re-emits the light. The shell layer is responsible to confine the emission and passivate defects in the structure. The ligand layer provides more stability. The addition of barrier layers is required to protect QDs from oxygen, water and heat. Quantum dot-based light emitting diodes (QDLEDs) are a new form of light emitting technology based on nanoparticle, and their structures are similar to the OLED technology. Although, in this technology, a layer of quantum dots is placed between electron and hole-transporting layers, like sandwiched structure. Electrons and holes are accumulated in the quantum dot layer by an applied electric field. Then, they will recombine and emit narrow spectrum of photons. For example, FWHM for Cd based is 25–35 nm and 40–50 nm for Cd-free QDs.

The efficiency of QDLEDs is still lower than OLEDs. But the pure emission colour, the easier tenability of colour emission by adjusting the particle size and their lower emitter cost make them interesting subject for researchers as well as artisans. Conducted researches improve the quantum efficiency of QDLEDs more than two order of magnitude up to now. A bounded electron and hole inside the QD can recombine and emit a photon that has energy equal to the gap between the highest occupied and lowest unoccupied states. In 1994, the first structure of organic light emitting diodes based on quantum dots is studied. This structure consists of a layer of CdSe quantum dots and the polymeric electron transport layer, which are placed between two electrodes [43]. Due to low mobility of organic semiconductor, QDLED had low performance, and the threshold voltage was as large as 4 V. Recently, a new colloidal quantum dot-based light emitting diode (QDLED) is reported with improved external quantum efficiencies (EQE) by applying the organic CIM/LiF/Al cathode [44]. QLEDs with this new structure increase the EQE about 25% comparing to the bare Al devices. Therefore, using an organic cathode interfacial material can result in better device performance, including the brightness, EQE and CE. In this proposed device, the peaks of EQE and CE were 8.5% and over 29 cd/A, respectively. This improvement is because of balanced electron/hole injection due to the presence of the organic CIM. The balancing of the carriers is hard, because most quantum dots are considered in n-type materials. So the current efficiency will be low in these devices. The p-type conductivity and hole injection barriers of the organic hole transport layer are necessary to improve the efficiency of QDLEDs. Further attempts are aimed to optimize charge injection, to transport, to improve stability of material and to control chemical and physical phenomena at the interface. Also, an all solution-processed QDLED with an inverted structure is investigated by Castan and his coworkers [45]. They demonstrated that the optimized amount of PTE in the PEDOT:PSS can balance the charge in the device. The red, green and blue devices using this structure have maximum luminance about 12.510, 32.370 and 249 cd/m^2 and turn-on voltages of 2.8, 3.6 and 3.6 V, respectively. Because of the process used for the fabrication of this device, it is very promising in the future of display industry.

In addition, highly bright and efficient blue QDLEDs have been reported by employing ZnCdSe core/multishell QDs as emitters [46]. The efficiency and brightness were improved by doping poly vinyl(N-carbazole) (PVK) in the emissive layer. It balances the charge injection because of the lower HOMO level, which causes the reduction of potential barrier at the interface of QDs and hole transport layer. This blue QDLEDs show a high efficiency (EQE > 8%), and the peak of efficiency happens at the luminance about 1000 cd/m^2. In 2007, Xie et al. found that the inorganic core oxidizes through their lifetime. So they suggested growing the shell materials on the surface of the core to passivate the inorganic core [47, 48]. They could improve stability by growing a ZnS shell around the InP core. The properties of QDs strongly depend on their shell and their compounds. The cluster diameter is a significant factor in determination of bandgap in structure of the QDs. The emission band will be narrower, while the diameter of the cluster gets smaller. Also the thickness of shell is important in increasing the maximum amount of the PL efficiency. According to the result of Bera and his colleagues' research, the thicker shell layer, the lower amount of photoluminescence quantum yield (PLQY) [49]. The main reason of this phenomenon is that the misfit dislocations (sites of non-radiative recombination) are formed when the shell layer is thick. Higher quantum efficiency will be available by minimizing these sites in QDs. In addition, matching the energy levels of the shell and core should be considered. Confining the excitons within the QDs is possible by selecting proper material of shell with wider bandgap to create an appropriate potential barrier around the QD.

Colloidal CdSe/ZnS (core-shell QDs) have high quantum yield and high photo stability at room temperature. So they are good choice in lighting industry, and many researchers have investigated them [50]. To prevent the light scattering, the particle size should be smaller than one-tenth of the visible light's wavelength [51]. On the other hand, large particles tend to accumulate that tarnish the composite film. The fluorescence properties of the QDs can be affected by the ability of QDs to aggregation. Accumulation effect can drastically reduce the quantum efficiency. Recent researches have demonstrated that the repulsive force between the molecular chains of polymers can prevent the accumulation of nanoparticles, so the compound of the polymer quantum dot can improve this problem. Up to now colloidal nanoparticles of cadmium sulphides, cadmium silicon and lead sulphides are used in organic light emitting diodes. These QDLEDs emit green light potentially [52–56]. On the other hand, these kinds of QDs have toxicological properties so it is environmentally restricted and not to be able to be a commercial material in this field. New Cd-free quantum dots should be introduced to commercialize the QDLED technology. ZnO cores with a MgO shell, InP-based dots and CuInS2 are three new materials that need more studies to be performed by scientists [57, 58]. In 2015, Du et al. studied a stable photoluminescence QDLEDs based on hydrophilic CdTe QD. Inorganic nanocomposite CdTe quantum dots were prepared with two rotary steam and freeze-drying methods. Because of adhesion, flexibility and transparency, silica gel can be coated on the surface of UV light emitting diode and form photoluminescence QDLEDs. This new photoluminescence QDLED is sustainable and cost-effective. Also it is easy to operate and environmentally non-toxic [59]. Recently, Kim' group has studied a multiple structure of QDLED based on InP quantum dots. Current efficiency and brightness in this structure are reported to be 1 cd/A and 530 cd/m^2, respectively. As mentioned, the best results in improving the stability of organic light emitting diodes are based on Cd QDs. InP quantum dots are replaced with Cd QDs in this study because of the environmental risks of cadmium [60]. In addition, the interface trap states are very effective on the performance of the device. In 2016, Koh et al. investigated these traps in the presence of TCNQ between charge transfer layer and quantum dots. With the introduction of TCNQ, the electroluminescent efficiency (EL) in QDLED has been improved by increasing the charge injection into the QD layer [61].

ZnO nanowires are perfect single crystals, which increases the mean free path of carriers transmitted on them. High density of QDs layers is prepared by ZnO nanowires which caused much more brightness of QDLEDs. These ZnO nanorods allow high density for QD and provide brighter LED-based display. This structure emits light that is very similar to the sun's light, and energy transfer efficiency of this structure is measured and equal to 17% [62]. The efficiencies equal to 10.7 and 14.5% are obtained by combining electron transfer layer and hole transfer layer and using the Cd QD for blue and green QDLED, respectively. However, the efficiency of red QDLED has reached to 20% [63]. Furthermore, the red QDLED with reverse multiple structures and excellent performance with an external quantum efficiency of 18% has been reported [64]. Up to now the external quantum efficiencies of QDLEDs based on Cd QD obtained are 10.7% [65] and 14.5% [66] for blue and green QDLEDs, respectively, while red QDLED efficiency is 20.5% [67]. Of course, the higher efficiencies are obtained in OLEDs without the use of quantum dots. In many applications, nanoparticles are imported into the polymer to give a special feature. For example, nanoparticles improve stability of the host material, because they act as energy absorbers to reduce the structural defects of organic materials. The benefits of using nanoparticles are high stability, narrow emission spectrum and feasible use in the polymer structure and the formation of thin film layers. Dark details, image sticking, peak luminance, colour gamut, colour volume, efficiency and lifetime of QDLEDs are so much better than OLEDs. However, black level, haloing, viewing angle and being eco-friendly are the advantages of OLEDs comparing to QDLEDs. QDLEDs are the energy efficient and have tuneable colour display. They deliver about 35% more luminous efficiency in comparison with OLEDs at the same colour point. Also, power efficiency of QDLEDs can be twice more than OLEDs at the same colour purity. The last but not least advantage of QDLEDs over OLEDs is low-cost manufacture. They can be printed in large area on thin flexible substrates, and they are also solution processable [68]. QDs have very narrow emission spectra, but their absorption spectra are broad. Factually, they absorb all wavelengths higher than their bandgap and convert them into a single colour. This narrow spectrum will improve colour saturation in QDLEDs compared to OLEDs. In addition, QDLEDs can be more power efficient due to good colour coordinate and luminous efficiency.

QDs can be used in solar cells as well as LEDs due to their broad excitation band and narrow emission spectra. The tunable colour of QDLEDs will be provided by controlling the quantum dot size [69]. For example, cadmium selenide quantum dots can emit optical wavelength in the range of 470–640 nm by varying the size of 2–8 nanometres. The size of the QD can make unique physical properties in QDLEDs because the electrons in a nanocrystal exhibit quantum mechanical effects. The quantum confinement phenomenon occurred in nanocrystal will lead to discrete energy levels. The bandgap energy of a QD is inversely proportional to its size; therefore the emission from a QD will be colour tuneable. At present, the best OLEDs can have a quantum efficiency of up to 33%, which is much higher than that of QDLEDs [70]. Defects in the crystal create some non-radiative electron-hole recombinations that are the main reason of the low quantum efficiency. Although, the PL efficiencies of QDs are high, still the EQEs in these devices are low mainly due to poor charge carrier injection into the QD layers [71]. QDLEDs will be a good choice for the future of LEDs due to their colour stability, easily tunable colour and long lifetime. In recent years, due to all the advantages of the QDLEDs mentioned above, many research groups have worked on QDLEDs [72–75], and the efficiency of this type of light emitting diodes has improved in subsequent researches [76–80].

9. Features of QDLEDs

QDLEDs are characterized by their total width at half maximum (FWHM). Moreover, having a high quantum yield and high charge transfer coefficient are two important features of the emissive layer [81]. FWHM is examined in these devices, and entirely these structures have a small FWHM. This value of a single QD size should be very small. However, the extension of the FWHM is unavoidable because there will be different sizes of the QDs. Because the size of the nanoparticles determines the wavelength of radiation and particles with similar size will be commensurate with the radiation intensity, radiation spectrum shows QD size distribution directly [82]. As mentioned above, the increase in FWHM shows that there is more diversity of QDs that can be caused by the reformation of QD result in exposure to UV and heat. When photons of UV are absorbed by colloidal quantum dots, the heat caused by losses stoke remains near to QDs and resizes QDs. Changes of FWHM will be more in higher currents. The use of semiconductor nanoparticles with narrow size distribution and narrow-band radiation leads to emit white light with low CRI. And this is because the CRI depends on the size and distribution of colloidal nanoparticles. In this way, we have developed a procedure for preparation of CdS colloidal nanocrystals. The emission spectrum of synthesized sample was shown in **Figure 4**. As can be found from this figure, FWHM of emission spectrum is reduced

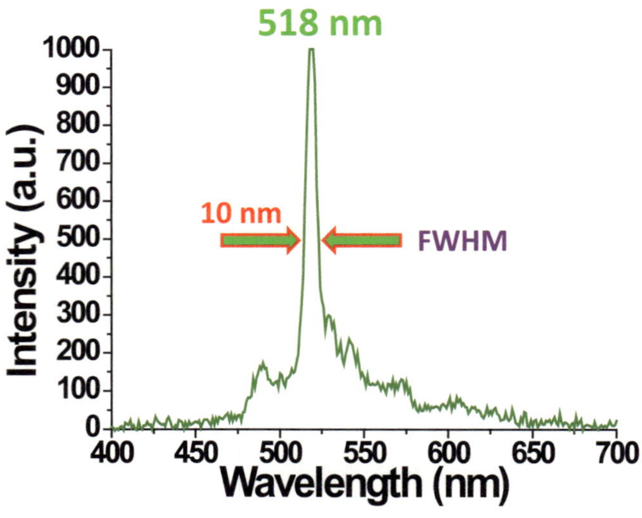

Figure 4. Emission spectrum of CdS colloidal sample.

to 10 nm, which is very small [83]. The prepared sample displays a strong and narrow green emission peak centred at 519 nm that has not been reported before, and it is longer than the onset of absorption of ~512 nm for bulk CdS. Several weak emission peaks appeared at wavelengths 490, 506, 521 and 543 nm, too. These two important characteristics of the prepared sample are due to the strong band-edge emission of CdS nanocrystals. **Figure 4** shows the PL spectrum of CdS nanoparticles excited by wavelength of 190 nm.

10. Different types of QDLEDs

QDs are applied in three types of OLEDs. PLEDs which their emissive layer is based on polymers, fluorescent small molecules and PHOLEDs which are the organo–metallic phosphorescent small molecules. Phase separation and contact printing are two major fabrication techniques for manufacturing of QDLED. **Table 1** shows common materials and QDs used in QDLEDs and OLEDs. Emission wavelength of QDs can be controlled by its size or composition.

QDs are very impressed by the environment (humidity and oxygen), because of the small size of the QDs. As can found from **Table 1**, ZnO is one of the semiconductors that can be used as electron transport material. Recently, we have developed a procedure for preparation of high mobility nanostructured thin indium-doped ZnO film [84, 85]. **Figure 5(a)** shows the scanning electron microscopy (SEM) of nanostructured thin ZnO film, and the X-ray diffraction (XRD) has been depicted in **Figure 5(b)**. It can be seen that there are three sharp diffraction peaks approximately at 30°, 33° and 35° that correspond to (1 0 0), (1 0 1) and (0 0 2).

Materials	HOMO	LUMO
PEDOT:PSS	−5.4	−2.4
MoO_3	−9.5	−6.5
a-NPB	−5.5	−2.4
ZnO	−7.5	−4
TPBI	−6.3	−2.8
TAZ	−6.4	−2.8
p-NiO	−5.4	−1.8
CdSe/ZnS (green QD)	−6.8	−4.3
CdSe/ZnS (red QD)	−6.7	−4.7
InP/ZnS (QD)	−5.2	−2.2

Table 1. Energy levels of some common hole and electron transport materials used in OLEDs and typical QDs.

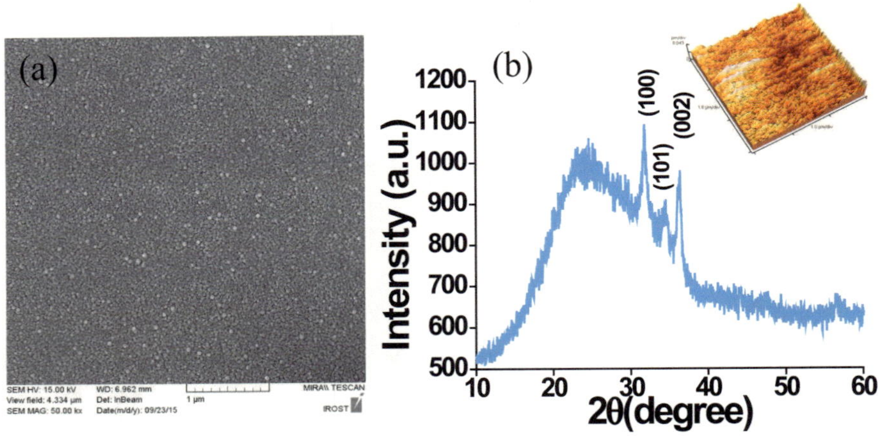

Used precursor	Band gap (eV)	Grain size (nm) (AFM)	Grain size (nm) (FESEM)	RMS (nm)	Resistivity (Ω.cm)	Carrier concentration n (cm^{-3})	Mobility μ(cm^2/V_s)
IZO	3.27	38	28	4.1	2.2 x 10^{-3}	4.4 x 10^{20}	6.49

Figure 5. (a) A scanning electron microscopy (SEM) image and (b) the X-ray diffraction (XRD) pattern of QD.

11. Conclusions and challenge ahead

It is expected to see much more progresses in the lighting industry particularly QDLEDs in the near future. Optimizing the efficiency of devices can help to improve the performance of QDLEDs. Many researchers try to do their best in this respect such as Shen and his group [86]. They have suggested a new high efficiency QDLED. Anikeeva et al. try to increase efficiency by using materials with high PLs in the red, green and blue regions of the visible spectrum [87]. Despite all efforts they made, improving the efficiency of blue QDLEDs seems to be challenging because of that the blue QDs and used electron and hole-transporting materials have low spectral overlap with each other. They expect that using wide bandgap hole and electron transporting organic materials improves the efficiency of the blue QDLEDs due to better exciton energy transfer and direct charge injection into the blue QDs. Hybrid devices that incorporate emissive layers using different types of emissive materials can play a big role in the future of QDLEDs. They could be made by a blue emitting TADF layer, a green phosphorescent layer and a red QD layer.

In conclusion, to improve the performance of the QDLEDs:

1. Improve the structure of QDLEDs.
2. Improve manufacturing techniques.
3. Choose a suitable material for the injection and transfer layers.
4. Structural differences of quantum dots.

There are a number of requirements that must be met in order for quantum dots to be integrated into the LED device and replace phosphor-based solutions. For one, the quantum dots should be stable in air, and moisture and the colour performance must be stable. Another problem faced to the development of the quantum dot materials is self-quenching. The quantum dots are designed to absorb light in one wavelength range and re-emit in another. The efforts of the researchers to create such a display are still in progress. QDLEDs promise to introduce very high contrast device, but with lower power than other technologies existed up to now. In addition, the lifetime of QDLEDs is another feature that needs more attention.

Acknowledgements

This chapter was supported by a National Research Foundation of Korea (NRF) grant funded by the South Korean government (MSIP 2015R1A2A1A15054541) and by the third stage of Brain Korea 21 Plus Project in 2016.

Author details

Neda Heydari[1]*, Seyed Mohammad Bagher Ghorashi[2], Wooje Han[3] and Hyung-Ho Park[3]*

*Address all correspondence to: nheydari.ph88@gmail.com and hhpark@yonsei.ac.kr

1 Institute of Nanoscience and Nanotechnology, University of Kashan, Kashan, Iran

2 Department of Physics, Faculty of Physics, University of Kashan, Kashan, Iran

3 Department of Materials Science and Engineering, Yonsei University, Seodaemun-gu, Seoul, Korea

References

[1] Kim SK, Chung TG, Chung DH, Lee HS, Song MJ, Park JW, ..., Kim TW. Improvement of efficiency in organic light-emitting diodes using PVK and CuPc buffer layer. Optical Materials. 2003;**21**(1):159-164

[2] Wen L, et al. Electroplex emission at PVK/Bphen interface for application in white organic light-emitting diodes. Journal of Luminescence. 2011;**131**(11):2252-2254

[3] Jeong SM, Lee DY, Koo WH, Choi SH, Baik HK, Lee SJ, Song KM. Improved stability of organic light-emitting diode with aluminum cathodes prepared by ion beam assisted deposition. Science and Technology of Advanced Materials. 2005;**6**(1):97-102

[4] Huang BJ, Tang CW. Thermal–electrical–luminous model of multi-chip polychromatic LED luminaire. Applied Thermal Engineering. 2009;**29**(16):3366-3373

[5] Tang CW, VanSlyke SA. Organic electroluminescent diodes. Applied Physics Letters. 1987;**51**(12):913-915

[6] Bing LY. On thermal structure optimization of a power LED lighting. Procedia Engineering. 2012;**29**:2765-2769

[7] Vijay SH A Healthy Future Forecast for the Outdoor LED Luminaire Lighting Market, Strategies unlimited; September 2016

[8] Bardsley N. Solid-State Lighting Research and Development: Multi-Year Program Plan provided for Lighting Research and Development Building Technologies Program: Office of Energy Efficiency and Renewable Energy, U.S. Department of Energy; April 2012

[9] Meadows C. LEDs magazine special report: SIL and the LED show bring out the social (media) butterflies. LEDs Magazine. March 2015

[10] Li Y, Rizzo A, Cingolani R, Gigli G. White-light-emitting diodes using semiconductor nanocrystals. Microchimica Acta. 2007;**159**(3-4): 207-215

[11] Tang CW, VanSlyke SA, Chen CH. Electroluminescence of doped organic thin films. Journal of Applied Physics. 1989;**65**(9):3610-3616

[12] Zhen-Gang L, Zhi-Jian C, Qi-Huang G. Reduction of concentration quenching in a non-doped DCM organic light-emitting diode. Chinese Physics Letters. 2005;**22**(6):1536

[13] Mizoguchi SK, Santos G, Andrade AM, Fonseca FJ, Pereira L, Iha NYM. Luminous efficiency enhancement of PVK based OLEDs with fac-[ClRe (CO) 3 (bpy)]. Synthetic Metals. 2011;**161**(17):1972-1975

[14] Kwong CY, et al. Efficiency and stability of different tris(8-hydroxyquinoline) aluminum (Alq3) derivatives in OLED applications. Material Science and Engineering B. 2005;**116**(1):75-81

[15] Rosselli FP, et al. Experimental and theoretical investigation of tris-(8-hydroxy-quinolinate) aluminum (Alq3) photo degradation. Organic Electronics. 2009;**10**(8):1417-1423

[16] Gebeyehu D. Highly efficient p-i-n type organic light-emitting diodes using doping of the transport and emission layers. Ethiopian Journal of Science and Technology. 2014;**7**(1):37-48

[17] Di D, Yang L, Richter JM, Meraldi L, Altamimi RM, Alyamani AY, … Friend RH. Efficient triplet exciton fusion in molecularly doped polymer light-emitting diodes. Advanced Materials. 2017;**29**(13):1605987

[18] Santos ER, Moraes JIBD, Takahashi CM, Sonnenberg V, Burini EC, Yoshida S., … Hui WS. Low cost UV-Ozone reactor mounted for treatment of electrode anodes used in P-OLEDs devices. Polímeros, (AHEAD). 2016;**26**(3):236-241

[19] Ho S, Chen Y, Liu S, Peng C, Zhao D, So F. Interface effect on efficiency loss in organic light emitting diodes with solution processed emitting layers. Advanced Materials Interfaces. 2016;**3**(19):1600320

[20] Lewis J, Material challenge for flexible organic devices. Materials Today. 2006;**9**(4):38-45

[21] Wu Z. et al. Transparent, conductive carbon nanotube films. Science. 2004;**305**(5688): 1273-1276

[22] van de Lagemaat J, et al. Organic solar cells with carbon nanotubes replacing In2O3:Sn as the transparent electrode, Applied Physics Letters. 2006;**88**(23):233503

[23] Bae S, et al. Roll-to-roll production of 30-inch graphene films for transparent electrodes. Nature Nanotechnology. 2010;**5**(8):574-578

[24] Han T-H, et al. Extremely efficient flexible organic light-emitting diodes with modified graphene anode. Nature Photonics. 2012;**6**(2):105-110

[25] Kang M-G, Guo LJ. Nanoimprinted semitransparent metal electrodes and their application in organic light-emitting diodes, Advanced Materials. 2007;**19**(10):1391-1396

[26] Kang M-G, Kim M-S, Kim JS, Guo LJ. Organic solar cells using nanoimprinted transparent metal electrodes, Advanced Materials. 2008;**20**(23):4408-4413

[27] Leftheriotis G, Yianoulis P, Patrikios D. Deposition and optical properties of optimised ZnS/Ag/ZnS thin films for energy saving applications. Thin Solid Films. 1997;**306**(1):92-99

[28] Liu X, Cai X, Mao J, Jin C. ZnS/Ag/ZnS nano-multilayer films for transparent electrodes in flat display application. Applied Surface Science. 2001;**183**(1):103-110

[29] Lee JY, Connor ST, Cui Y, Peumans P. Solution-processed metal nanowire mesh transparent electrodes. Nano Letters. 2008;**8**(2):689-692

[30] Azulai D, Belenkova T, Gilon H, Barkay Z, Markovich G. Transparent metal nanowire thin films prepared in mesostructured templates, Nano Letters. 2009;**9**(12):4246-4249

[31] Burns S, MacLeod J, Do TT, Sonar P, Yambem SD. Effect of thermal annealing super yellow emissive layer on efficiency of OLEDs. Scientific Reports. 2017;**7**:40805

[32] Yersin H, Rausch AF, Czerwieniec R. Organometallic emitters for OLEDs: Triplet harvesting, singlet harvesting, case structures, and trends. Physics of Organic Semiconductors. 2nd ed. 2012:371-424

[33] Tanaka H, Shizu K, Miyazaki H, Adachi C. Efficient green thermally activated delayed fluorescence (TADF) from a phenoxazine–triphenyltriazine (PXZ–TRZ) derivative. Chemical Communications. 2012;**48**(93):11392-11394

[34] Kim M, Jeon SK, Hwang SH, Lee SS, Yu E, Lee JY. Highly efficient and color tunable thermally activated delayed fluorescent emitters using a "twin emitter" molecular design. Chemical Communications. 2016;**52**(2):339-342

[35] Santos PL, Ward JS, Data P, Batsanov AS, Bryce MR, Dias FB, Monkman AP. Engineering the singlet–triplet energy splitting in a TADF molecule. Journal of Materials Chemistry C. 2016;**4**(17):3815-3824

[36] Sagara Y, Shizu K, Tanaka H, Miyazaki H, Goushi K, Kaji H, Adachi C. Highly efficient thermally activated delayed fluorescence emitters with a small singlet–triplet energy gap and large oscillator strength. Chemistry Letters. 2014;**44**(3):360-362

[37] Evans RC, Douglas P, Winscom, CJ. Coordination complexes exhibiting room-temperature phosphorescence: Evaluation of their suitability as triplet emitters in organic light emitting diodes. Coordination Chemistry Reviews. 2006;**250**(15):2093-2126

[38] Geffroy B, Le Roy P, Prat C. Organic light-emitting diode (OLED) technology: Materials, devices and display technologies. Polymer International. 2006;**55**(6):572-582

[39] Kietzke T. Recent advances in organic solar cells. Advances in OptoElectronics. 2008;**2007**:40285

[40] Kim KH, Lee S, Moon CK, Kim SY, Park YS, Lee JH, ... Kim JJ. Phosphorescent dye-based supramolecules for high-efficiency organic light-emitting diodes. Nature Communications. 2014;**5**:4769

[41] McElvain J, Antoniadis H, Hueschen MR, Miller JN, Roitman DM, Sheats JR, Moon RL. Formation and growth of black spots in organic light-emitting diodes. Journal of Applied Physics. 1996;**80**(10):6002-6007

[42] Kickelbick G. Concepts for the incorporation of inorganic building blocks into organic polymers on a nanoscale. Progress in Polymer Science. 2003;**28**(1):83-114

[43] Colvin VL, Schlamp MC, Paul Alivisatos A. Light-emitting diodes made from cadmium selenide nanocrystals and a semiconducting polymer. 1994:354-357

[44] Ding T, Yang X, Ke L, Liu Y, Tan WY, Wang N, ... Sun XW. Improved quantum dot light-emitting diodes with a cathode interfacial layer. Organic Electronics. 2016;**32**:89-93

[45] Castan A, Kim HM, Jang J. All-solution-processed inverted quantum-dot light-emitting diodes. ACS Applied Materials & Interfaces. 2014;**6**(4):2508-2515

[46] Wang L, Chen T, Lin Q, Shen H, Wang A, Wang H, ... Li LS. High-performance azure blue quantum dot light-emitting diodes via doping PVK in emitting layer. Organic Electronics. 2016;**37**:280-286

[47] Xie R, Battaglia D, Peng X. Colloidal InP nanocrystals as efficient emitters covering blue to near-infrared. Journal of the American Chemical Society. 2007;**129**(50):15432-15433

[48] Lim J, Bae WK, Lee D, Nam MK, Jung J, Lee C, ... Lee S. InP@ ZnSeS, core@ composition gradient shell quantum dots with enhanced stability. Chemistry of Materials. 2011;**23**(20):4459-4463

[49] Bera D, Qian L, Tseng TK, Holloway PH. Quantum dots and their multimodal applications: A review. Materials. 2010;**3**(4):2260-2345

[50] Nguyen HT, Pham TN, Koh KH, Lee S. Fabrication and characterization of CdSe/ZnS quantum-dot LEDs. Physic Status Solidi (a). 2012;**209**(6):1163-1167

[51] Althues H, Henle J, Kaskel S. Functional inorganic nanofillers for transparent polymers. Chemical Society Reviews. 2007;**36**(9):1454-1465

[52] Ghosh B, Sakka Y, Shirahata N. Efficient green-luminescent germanium nanocrystals. Journal of Materials Chemistry A. 2013;**1**(11):3747-3751

[53] Hua F, Swihart MT, Ruckenstein E. Efficient surface grafting of luminescent silicon quantum dots by photoinitiated hydrosilylation. Langmuir. 2005;**21**(13):6054-6062

[54] Baker DR, Kamat PV. Tuning the emission of CdSe quantum dots by controlled trap enhancement. Langmuir. 2010;**26**(13):11272-11276

[55] Ghamsari MS, Ara MM, Radiman S, Zhang XH. Colloidal lead sulfide nanocrystals with strong green emission. Journal of Luminescence. 2013;**137**:241-244

[56] Efafi B, Ghamsari MS, Aberoumand MA, Ara MM, Rad HH. Highly concentrated ZnO sol with ultra-strong green emission. Materials Letters. 2013;**111**:78-80

[57] Kim HM, Jang J. High-efficiency inverted quantum-dot light emitting diodes for display. In SID Symposium Digest of Technical Papers. Vol. 45, No. 1; 2014, June. pp. 67-70

[58] Chen B, Zhong H, Zhang W, Tan ZA, Li Y, Yu C, … Zou B. Highly emissive and color-tunable CuInS2-based colloidal semiconductor nanocrystals: Off-stoichiometry effects and improved electroluminescence performance. Advanced Functional Materials. 2012;**22**(10):2081-2088

[59] Du J, Wang C, Xu X, Wang Z, Xu S, Cui Y. Assembly of light-emitting diode based on hydrophilic CdTe quantum dots incorporating dehydrated silica gel. Luminescence. 2016;**31**(2):419-422

[60] Kim HY, Park YJ, Kim J, Han CJ, Lee J, Kim Y, … Oh MS. Transparent InP quantum dot light-emitting diodes with ZrO2 electron transport layer and indium zinc oxide top electrode. Advanced Functional Materials. 2016;**26**(20):3454-3461

[61] Koh WK, Shin T, Jung C, Cho DKS. TCNQ interlayers for colloidal quantum dot light-emitting diodes. ChemPhysChem. 2016;**17**(8):1095-1097

[62] Zhao X, Liu W, Chen R, Gao Y, Zhu B, Demir HV, … Sun H. Exciton energy recycling from ZnO defect levels: towards electrically driven hybrid quantum-dot white light-emitting-diodes. Nanoscale. 2016;**8**(11):5835-5841

[63] Kim JH, Jo DY, Lee KH, Jang EP, Han CY, Jo JH, Yang H. White electroluminescent lighting device based on a single quantum dot emitter. Advanced Materials. 2016;**28** (25):5093-5098

[64] Mashford BS, Stevenson M, Popovic Z, Hamilton C, Zhou Z, Breen C, … Kazlas PT. High-efficiency quantum-dot light-emitting devices with enhanced charge injection. Nature Photonics. 2013;**7**(5):407-412

[65] Shen H, Cao W, Shewmon NT, Yang C, Li LS, Xue J. High-efficiency, low turn-on voltage blue-violet quantum-dot-based light-emitting diodes. Nano Letters. 2015;**15**(2):1211-1216

[66] Yang Y, Zheng Y, Cao W, Titov A, Hyvonen J, Manders JR, … Qian L. High-efficiency light-emitting devices based on quantum dots with tailored nanostructures. Nature Photonics. 2015;**9**:259-266

[67] Dai X, Zhang Z, Jin Y, Niu Y, Cao H, Liang X, … Peng X. Solution-processed, high-performance light-emitting diodes based on quantum dots. Nature. 2014:**515**(7525):96-99

[68] Anikeeva PO, Madigan CF, Coe-Sullivan SA, Steckel JS, Bawendi MG, Bulović V. Photoluminescence of CdSe/ZnS core/shell quantum dots enhanced by energy transfer from a phosphorescent donor. Chemical Physics Letters. 2006;**424**(1):120-125

[69] Schreuder MA, Xiao K, Ivanov IN, Weiss SM, Rosenthal SJ. White light-emitting diodes based on ultrasmall CdSe nanocrystal electroluminescence. Nano Letters. 2010;**10**(2):573-576

[70] Chang HW, Lee J, Hofmann S, Hyun Kim Y, Müller-Meskamp L, Lüssem B, … Gather MC. Nano-particle based scattering layers for optical efficiency enhancement of organic light-emitting diodes and organic solar cells. Journal of Applied Physics. 2013;**113**(20):204502

[71] Zyga L. Quantum dot LEDs get brighter, more efficient. Phys Org. April 2012;1-3

[72] Anikeeva PO, Halpert JE, Bawendi MG, Bulović V. Electroluminescence from a mixed red– green– blue colloidal quantum dot monolayer. Nano Letters. 2007;**7**(8):2196-2200

[73] Cho KS, Lee EK, Joo WJ, Jang E, Kim TH, Lee SJ, … Kim JM. High-performance cross-linked colloidal quantum-dot light-emitting diodes. Nature Photonics. 2009;**3**(6):341-345

[74] Coe S, Woo WK, Bawendi M, Bulović V. Electroluminescence from single monolayers of nanocrystals in molecular organic devices. Nature. 2002;**420**(6917):800-803

[75] Sun Q, Wang YA, Li LS, Wang D, Zhu T, Xu J, … Li Y. Bright, multicoloured light-emitting diodes based on quantum dots. Nature Photonics. 2007;**1**(12:717-722

[76] Caruge JM, Halpert JE, Bulović V, Bawendi MG. NiO as an inorganic hole-transporting layer in quantum-dot light-emitting devices. Nano Letters. 2006;**6**(12):2991-2994

[77] Zhao J, Bardecker JA, Munro AM, Liu MS, Niu Y, Ding IK, … Ginger DS. Efficient CdSe/CdS quantum dot light-emitting diodes using a thermally polymerized hole transport layer. Nano Letters. 2006;**6**(3):463-467

[78] Kamat PV. Boosting the efficiency of quantum dot sensitized solar cells through modulation of interfacial charge transfer. Accounts of Chemical Research. 2012;**45**(11):1906-1915

[79] Matras-Postolek K, Bogdal D. Polymer nanocomposites for electro-optics: Perspectives on processing technologies, material characterization, and future application. In Polymer Characterization. Berlin Heidelberg: Springer; 2010. 221-282

[80] Fojtik A, Henglein A. Surface chemistry of luminescent colloidal silicon nanoparticles. The Journal of Physical Chemistry B. 2006;**110**(5):1994-1998

[81] Wood V, Panzer MJ, Halpert JE, Caruge JM, Bawendi MG, Bulovic V. Selection of metal oxide charge transport layers for colloidal quantum dot LEDs. ACS Nano. 2009;**3**(11):3581-3586

[82] Hsu SC, Chen YH, Tu ZY, Han HV, Lin SL, Chen TM, … Lin CC. Highly stable and efficient hybrid quantum dot light-emitting diodes. IEEE Photonics Journal. 2015;**7**(5):1-10

[83] Sasani Ghamsari M, Sasani Ghamsari AH. CdS colloidal nanocrystals with narrow green emission. Journal of Nanophotonics. 2016;**10**(2):026007/doi: 10.1117/1.JNP.10.026007

[84] Alamdari S, Jafar Tafreshi M, Sasani Ghamsari M. The effects of Indium precursors on the structural, optical and electrical properties of nanostructured thin ZnO films. Material Letters. 2017;**197**(15):94-97

[85] Sasani Ghamsari M, Alamdari S, Han W, Park HH. Impact of nanostructured thin ZnO film in ultraviolet protection. International Journal of Nanomedicine. 2017;**12**:207

[86] Shen H, Lin Q, Wang H, Qian L, Yang Y, Titov A, … Li LS. Efficient and bright colloidal quantum dot light-emitting diodes via controlling the shell thickness of quantum dots. ACS Applied Materials & Interfaces. 2013;**5**(22):12011-12016

[87] Anikeeva PO, Halpert JE, Bawendi MG, Bulovic V. Quantum dot light-emitting devices with electroluminescence tunable over the entire visible spectrum. Nano Letters. 2009; **9**(7):2532-2536

Chapter 4

Perovskite Quantum Dot Light-Emitting Diodes

Zhifeng Shi, Xinjian Li and Chongxin Shan

Additional information is available at the end of the chapter

http://dx.doi.org/10.5772/intechopen.68275

Abstract

Recently, lead halide perovskite quantum dots (QDs) have attracted much attention because of their excellent properties of high color purity, tunable emission wavelength covering the whole visible region, and ultrahigh photoluminescence (PL) quantum yield. They are expected to be promising candidates for the next-generation cost-effective lighting and display sources. Here, we introduced the recent development in the direct solution-processed synthesis and ion exchange-based reactions, leading to organic/inorganic hybrid halide perovskites ($CH_3NH_3PbX_3$; X = Cl, Br, I) and all-inorganic lead halide perovskites ($CsPbX_3$; X = Cl, Br, I), and studied their optical properties related to exciton-related emission and quantum confinement effect. Finally, we reviewed the recent progresses on the perovskite light-emitting diodes (LEDs) based on $CH_3NH_3PbX_3$ and $CsPbX_3$ quantum dots and provided a critical outlook into the existing and future challenges.

Keywords: perovskite, $CH_3NH_3PbX_3$, cesium lead halide, light-emitting diodes

1. Introduction

Motivated by the remarkable color tunability and relatively high photoluminescence (PL) quantum yield of colloidal quantum dots (QDs), the concepts of QD-based light-emitting diodes (LEDs) have been proposed and developed for a few years, and multicolor LEDs were successfully fabricated either for PL or electroluminescence (EL) mechanism [1–6]. It has been recognized that using QD-based LEDs as the backlighting system of liquid crystal display can greatly expand the color gamut of display and present vibrant colored images [7]. Although the conventional CdSe-based QD system has been commercially used, it suffers from the lack of surface control during process and low-cost preparation technique [8]. Moreover, development of high-performance CdSe QD-based LEDs strongly relies on the precise core/shell design, involving the band engineering and surface ligands. In this circumstance, colloidal lead halide perovskite QDs began to attract a great scientific attention. The appealing properties of this new class of materials that might enable advances in electroluminescent devices are their outstanding optical properties including the narrow emission band (<20 nm), a wide wavelength tunability (400–800 nm), and a high PL quantum yield, which make them suitable for high-performance, low-cost, and lightweight LED applications [9–11]. Beyond LEDs, they were also explored as interesting materials for low-threshold lasing [12], photodetectors [13], and solar cells [14]. And, many studies on these fields have been reported recently. In the following, two typical synthesized methods for the novel perovskite QDs system were introduced in detail, and their optical properties related to the exciton-related emission and quantum confinement effect were investigated. Finally, we reviewed the previously reported device structures on perovskite QD LEDs and provided a critical outlook into the existing and future challenges.

2. Crystal structure features

Lead halide perovskite QDs have a crystal structure of ABX_3, in which A and B are the monovalent and divalent cations, respectively, and X is a monovalent halide anion (Cl, Br, I). The typical crystal structure of lead halide perovskites was illustrated in **Figure 1a**. B cation is coordinated to six halide ions in an octahedral configuration, and the octahedra are corner-sharing, with the A cation located in between those octahedra. In addition, the lead halide perovskite QDs can fall into two categories depending on the chemical component of A cation, organic-inorganic hybrid, and all-inorganic perovskites. Methylammonium (MA) lead trihalide perovskites have so far been the most intensively explored in optoelectronics, and they have the chemical composition of $MAPbX_mY_{3-m}$. This perovskite is a hybrid inorganic-organic direct-bandgap semiconductor. Also, the perovskites based on colloidal $CsPbX_3$ QDs are extensively investigated in recent years. As is well known, $MAPbX_3$ or $CsPbX_3$ QDs can crystalline in orthorhombic, tetragonal, and cubic polymorphs according to the environment temperature and total system energy. And, experimental results have confirmed that $MAPbX_3$ or $CsPbX_3$ QDs possess different crystal structures at different temperatures or with different halide ions [15]. For the

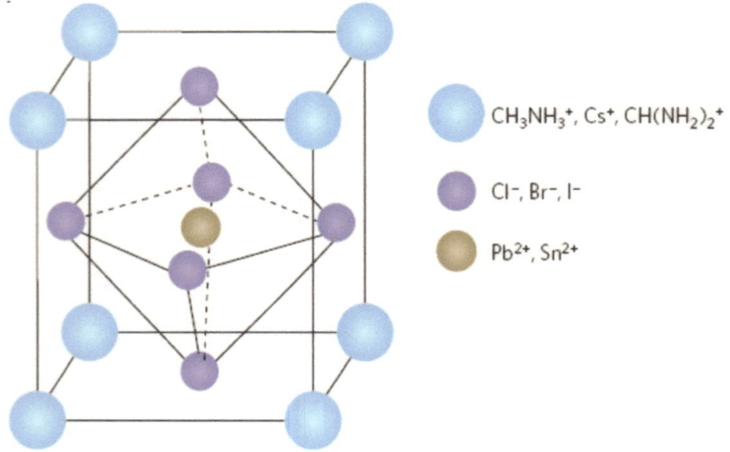

Figure 1. Illustration of the perovskite crystal structure.

two types of perovskite QDs, their optical and electronic properties are tunable by varying the composition of constituted halide ions and a smaller degree of the cations. Also, the size of perovskite QDs plays an important role on their optical properties due to quantum confinement effect [16].

3. Chemical synthesis of perovskite QDs

3.1. Synthesis of $MAPbX_3$ QDs

Perez-Prieto and coworkers pioneered the wet chemistry colloidal synthesis of free-standing hybrid perovskite QDs ($MAPbBr_3$ in their case) [17]. A simple one-step approach was employed in their experiment. The $PbBr_2$ was reacted directly with a mixture of ammonium bromide with short methyl chain and longer alkyl chains. In the middle part of the crystal, the MA cations were embedded, which would connect the neighboring $[PbBr_6]^-$ octahedra. And, the outer space will be occupied by the longer alkyl ammonium cations; as a result, the growth of $MAPbBr_3$ nanocrystals (NCs) is terminated. Therefore, one could assume that the longer alkyl ammonium cations in the outer space played the role of capping ligands of the perovskite NCs, which may be the reason that the $MAPbBr_3$ NCs could be dispersible in many solvents. That is to say that the role of these ligands is to provide a self-termination of the crystallization, leading to the formation of discrete nanoparticles in solution. In their case, the resulting $MAPbBr_3$ QDs could be stable for concentrated solutions as well as in solid states for 3 months. The corresponding microstructure of the synthesized products was shown in **Figure 2a**, in which a mixture of nanodots (~10 nm) and nanoplatelets (~40 nm) could be distinguished. **Figure 2b** and **c** shows the absorption and PL spectra of these highly crystalline $MAPbBr_3$ QDs, and a high-purity green emission at about 530 nm can be found. The corresponding PL quantum efficiency was about 20%.

By optimizing the molar ratio of octylammonium bromide:MA bromide:$PbBr_2$ (8:12:5) in a typical reprecipitation method while maintaining the 1-octadecene:$PbBr_2$ molar ratio of 62.6:1.0, this research group promoted the quantum efficiency of $MAPbBr_3$ QDs to 83% [18], demonstrating a promising potential for use in luminescent devices, such as LEDs and laser diodes. Owing to the fact that surface states in QDs would support the desired passivation treatment, Perez-Prieto and colleagues improved the organic capping of the $MAPbBr_3$ QDs; as a result, intensely luminescent and easily dispersible $MAPbBr_3$ QDs were produced, as shown in **Figure 2d**. As for the origin of the surface states, Li et al. recognized that the richness of halogen at the surface of QDs ought to be responsible for surface states [19]. On the one hand, abundant Br atoms at the surface will connect with cations, inhibiting the trapping of excited carriers and then high PL quantum efficiency. This process can be named as self-passivation effect, which is similar to the intentional passivation behaviors of traditional QDs with halogen ions. On the other hand, riched Br on surface should be binded with MA to form $PbBr_x$ analogs, which possesses a relatively large bandgap of 4.0 eV. Therefore, quasi-core-shell structure of $MAPbBr_3$/Br was formed. This superior characterization is similar as the shelling of ZnS around CdSe QDs core, photoexcitated carriers would be confined, and a high PL quantum efficiency could be expected. For a further confirmation, Zhong and coworkers performed the energy-dispersive spectroscopy measurements for synthesized $MAPbBr_3$ QDs [20]. They found a Br/Pb ratio of 3:5 for QDs with an average diameter of 3.3 nm, matching with the observation in Li's report. Also, the X-ray photoelectron spectroscopy results for $MAPbBr_3$ QDs support their argument above, and the smaller the diameter of QDs, the larger the ratio of Br/Pb.

Subsequently, this simple synthetic method was adopted in other groups, and an increasing number of papers have been published in this field. For example, Luo and coworkers used either octylammonium bromide or octadecylammonium bromide to produce perovskite QDs by using the reprecipitation method, with the size of QDs of 3.9 and 6.5 nm, respectively [21]. In their case, the size control of QDs was due to the different ligand-binding kinetics, and the high solubility of longer hydrophobic chain ligands could facilitate the increase of QD size. Zou et al. developed a ligand-assisted reprecipitation strategy to fabricate highly luminescent and color-tunable colloidal $MAPbBr_3$ QDs with the absolute quantum efficiency up to 70% at room temperature (RT) and low excitation fluencies. In their work, the reprecipitation method was a simple way for preparing organic nanocrystals or polymer dots simultaneously through the solvent mixing [22]. As shown in **Figure 3a**, they employed the same principle and simply mix a solution of $MAPbX_3$ precursors in good solvent (*N*-dimethylformamide (DMF)) into a vigorously stirred poor solvent (toluene, hexane, etc.) to form the organometal halide perovskites. Simultaneously, the long-chain organic ligands, such as n-octylamine and

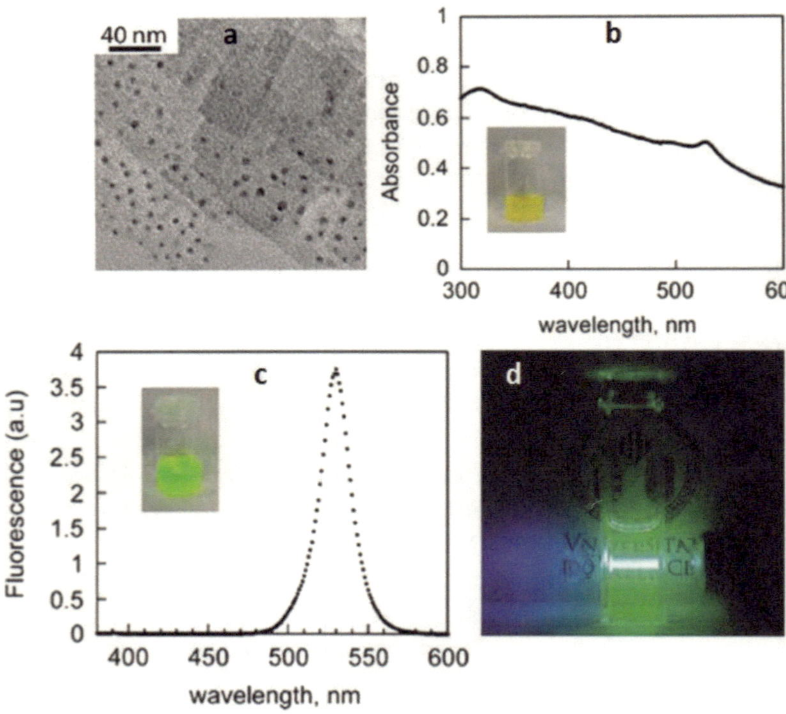

Figure 2. (a) Transmission electron microscope image of $MAPbBr_3$ QDs with a mixture of nanodots and nanoplatelets. (b) UV-visible absorption and (c) PL spectra of the $MAPbBr_3$ QDs. (d) Image of the toluene dispersion of $MAPbBr_3$ QDs under UV-laser pointer excitation.

oleic acid (OA), were introduced into a mixture to control the crystallization of precursors into colloidal QDs. So, it can be assumed that n-octylamine controlled the kinetics of crystallization and further contributed to the size control of $MAPbX_3$ QDs, whereas OA suppressed the aggregation effects and ensures their colloidal stability. **Figure 3b** shows the typical transmission electron microscopy image of $MAPbBr_3$ QDs, which were quasi-spherical and had an average diameter of 3.3 nm with a size deviation of ±0.7 nm. In addition, the simple ligand-assisted reprecipitation approach can be easily extended to fabricate colloidal $MAPbX_3$ QDs through halide substitutions. By mixing of PbX_2 slats in the precursors, a series of QDs with tunable compositions and emission wavelengths were prepared. **Figure 3c** shows the PL spectra and optical images of these samples under sunlight and an UV lamp (365 nm).

3.2. Synthesis of $CsPbX_3$ QDs

Colloidal synthesis routes of all-inorganic $CsPbX_3$ QDs have also been developed, and such novel QD systems have been receiving increasingly significant attention, which is assumably due to the reason that $CsPbX_3$ QDs possess higher stability than that of organic-inorganic hybrid $MAPbX_3$ QDs. Following the traditional hot-injection approach which is commonly used for the synthesis of metal chalcogenide QDs, Kovalenko and coworkers firstly synthesized the monodisperse $CsPbX_3$ QDs with a high degree of compositional bandgap engineering [23]. In their report, the Cs precursors were injected into the lead halide precursors, which contained the hot, high boiling point solvents. OA and oleylamine were mixed to dissolve the lead halide sources and to stabilize the QDs. By using the in situ PL measurements, they observed an interesting phenomenon that the reaction process after the Cs precursor injection was very quick. Within several seconds, the synthesis of the majority of $CsPbX_3$ QDs was realized. For the above synthesis method with a very rapid process, the unit size of $CsPbX_3$ QDs depends strongly on

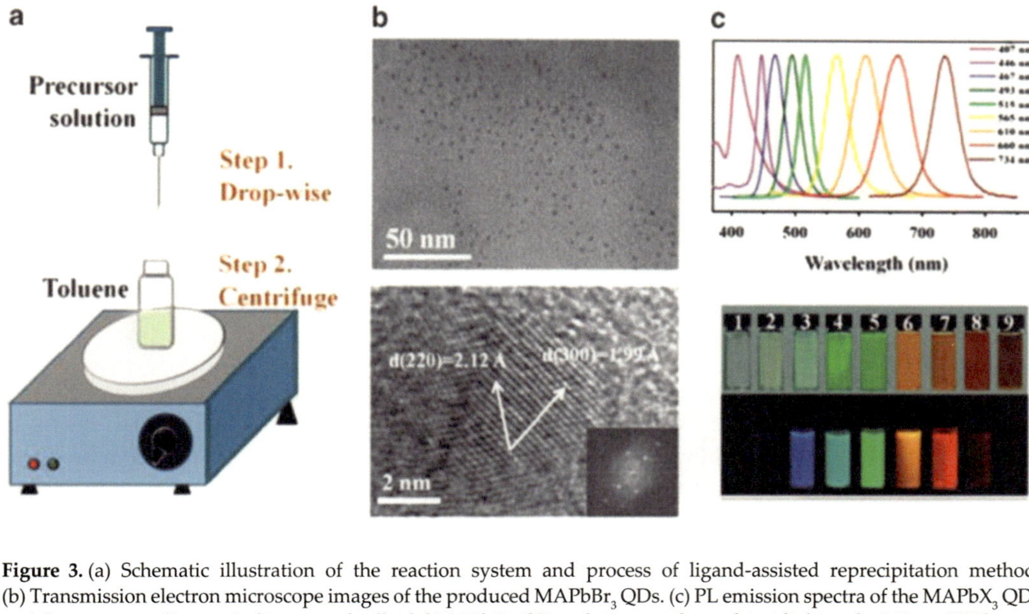

Figure 3. (a) Schematic illustration of the reaction system and process of ligand-assisted reprecipitation method. (b) Transmission electron microscope images of the produced $MAPbBr_3$ QDs. (c) PL emission spectra of the $MAPbX_3$ QDs and the corresponding optical images of colloidal $MAPbBr_3$ QDs solutions under ambient light and a 365 nm UV lamp.

the reaction temperature, and the size of the QDs decreases by decreasing the reaction temperature. In their study, the $CsPbX_3$ QDs were produced with tunable size from 4 to 15 nm. Note that $CsPbX_3$ QDs crystallized in the cubic phase in their case rather than the tetragonal or orthorhombic phases at high temperature, as shown in **Figure 4a** and **b**. More importantly, the emission wavelength or photon energy of the resulting $CsPbX_3$ QDs was tunable over the entire visible spectral region (410–700 nm) by varying the ratios of the precursor salts ($PbCl_2/PbBr_2$, $PbBr_2/PbI_2$). As shown in **Figure 4c** and **d**, the obtained $CsPbX_3$ QDs exhibit continuously tunable PL emission with a narrow linewidth of 12–42 nm and also a superior PL quantum yield up to 90%. Such an excellent optical performance is interesting, considering the facile synthesis route which involves neither core-shell structure nor surface modifications. Later on, by means of a droplet-based microfluidic platform, allowing for online absorption/PL measurements, the same research group investigated the formation mechanisms of such perovskite QDs.

In addition to the hot-injection technique, RT reprecipitation methods have also been proposed and developed for the controllable synthesis of $CsPbX_3$ QD system. For instance, Huang et al. reported the emulsion-based synthesis of perovskite nanocrystals involving many morphologies at RT [20]. Li et al. reported that the $CsPbX_3$ QDs can be synthesized at RT, which was similar to the ligand-assisted reprecipitation method used for $MAPbX_3$ QD system [19].

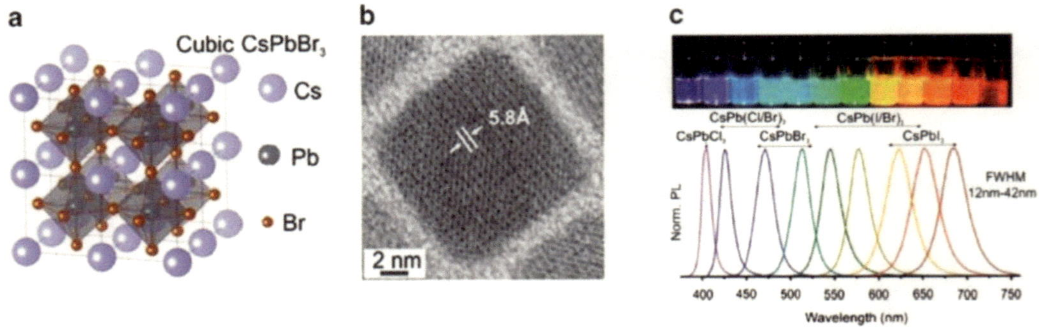

Figure 4. Monodisperse $CsPbX_3$ QDs and their structural characterization. (a) Schematic of the cubic perovskite lattice and (b) typical microstructure image of $CsPbX_3$ QDs. (c) Colloidal $CsPbX_3$ QDs exhibiting composition-tunable bandgap energies covering the entire visible region with narrow and bright emission (colloidal solutions in toluene under UV lamp).

4. Synthesis of perovskite QDs through halide ion exchange reaction

4.1. Synthesis of $MAPbX_3$ QDs

A post-preparative halide ion exchange on perovskite QDs provides an additional means to modify their composition and thus the optical characterizations while preserving their size and morphology. Jang et al. reported the reversible halide exchange reaction of organometal trihalide perovskite colloidal QDs for full-range bandgap tuning [24]. **Figure 5a** shows a schematic of the reversible halide exchange reaction of $MAPbX_3$ QDs with MAX, where X = Cl, Br, and I. The $MAPbX_3$ can be converted to any composition ones using MAX in isopropyl solution at RT. In their case, the synthesis of composition-tuned $MAPbBr_{3-x}Cl_x$ and $MAPbBr_{3-x}I_x$ was carried out by the Br exchange reaction of $MAPbBr_3$. As the starting material, $MAPbBr_3$ QDs were synthesized using a mixture of 1:1 MABr:$PbBr_2$ dissolved in octylamine and (octadecene) ODE. Octylamine serves as the capping ligands for the QDs. Then, $MAPbBr_3$ QDs were added into a MACl- or MAI-dissolved isopropyl solution affording $MAPbBr_{3-x}Cl_x$ and $MAPbBr_{3-x}I_x$, respectively. As shown in **Figure 5b**, the composition of the mixed halide perovskite QDs can be easily distinguished by their colors. **Figure 5c** shows the UV-visible diffuse reflectance spectrum of all produced samples (10 μm thick films on silicon substrates). The composition tuning of these samples enabled the bandgap to display absorption over a wide range of 400–850 nm, corresponding to the photon energy of 1.5–3.1 eV. The corresponding PL spectra of these samples were also measured and put together on a normalized scale for a comparison. Note that the emission intensity of $MAPbBr_{3-x}Cl_x$ and $MAPbBr_{3-x}I_x$ significantly decreases with increasing x, matching with other reports either for perovskite QDs or films. The anion-exchange method is also possible in the solid phase, as demonstrated by Yang and coworkers [25]. Following a solution-phase growth of $MAPbBr_3$ nanorod arrays, they were subsequently converted to $MAPbI_3$ with similar morphology via a low-temperature annealing at ~140°C in MAI vapor. Such an approach was further confirmed to be feasible by constructing heterostructured LEDs, with $MAPbBr_3$ nanorods exhibiting a green EL emission at 533 nm and $MAPbI_3$ nanorods emitting a red emission at 782 nm.

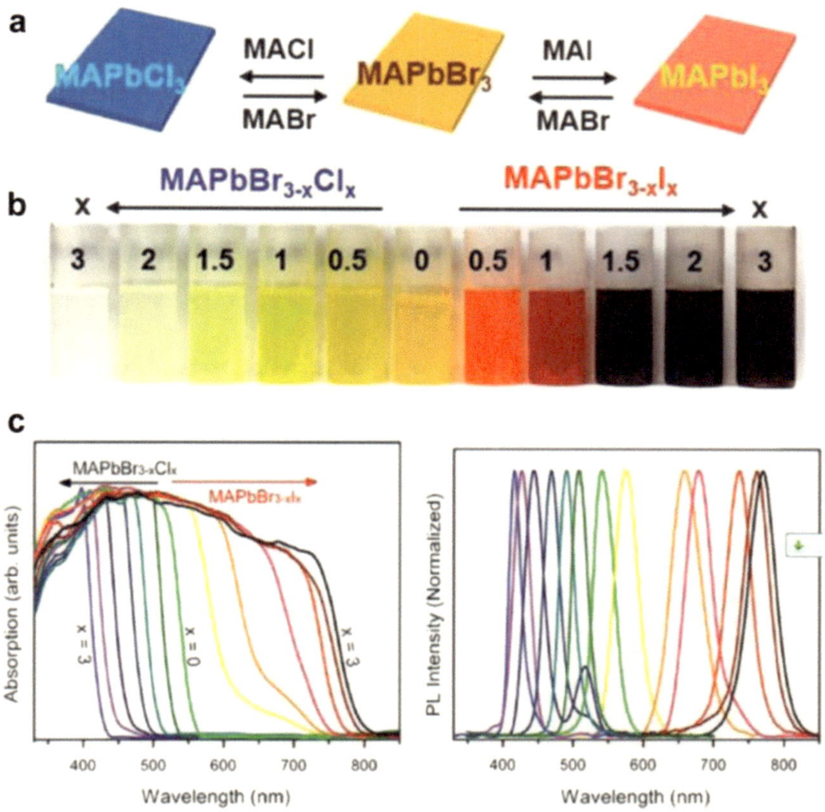

Figure 5. (a) Schematic of anion-exchange reactions for $MAPbX_3$ QD synthesis. (b) Photographs of mixed-halide $MAPbBr_{3-x}Cl_x$ and $MAPbBr_{3-x}I_x$ QDs under room light. (c) UV-visible absorption and RT PL spectra of $MAPbBr_{3-x}Cl_x$ and $MAPbBr_{3-x}I_x$ QDs.

4.2. Synthesis of $CsPbX_3$ QDs

As for the synthesis of $CsPbX_3$ QDs by halide ion exchange, Kovalenko and coworkers and Manna and coworkers have almost simultaneously reported this method for $CsPbX_3$ (X = Cl, Br, I) QD systems [26, 27], tuning their emission wavelength over the spectra range of 410–700 nm, as shown in **Figure 6a** and **b**. In their cases, five kinds of halide ion sources were investigated for ion exchange reaction, oleyammonium/octadecylammonium/tetrabutylammonium halides, organometallic Grignard reagents, and PbX_2 salts. An interesting observation was that the anion-exchange reaction between different halide ions is very fast, and the halide ions could achieve exchange reaction within seconds. A blueshift trend includes I^- to Br^- and Br^- to Cl^- routes, and, similarly, a redshift trend includes Cl^- to Br^- and Br^- to I^- routes, respectively. An important fact should be pointed out that a little size variation for $CsPbX_3$ QDs can be distinguished after the anion-exchange reaction, but the shape of the produced QDs was identical to their original appearance (parent QDs). Also, there was no change on the crystal structure of the produced QDs. Additionally, the ion exchange reaction process did not induce the undesired formation of any remarkable lattice and/or surface defects. As a result, the optical properties of anion-exchanged $CsPbX_3$ QDs are remarkable, including the linewidth and PL quantum yield, and comparable with those directly synthesized through the hot-injection method introduced above.

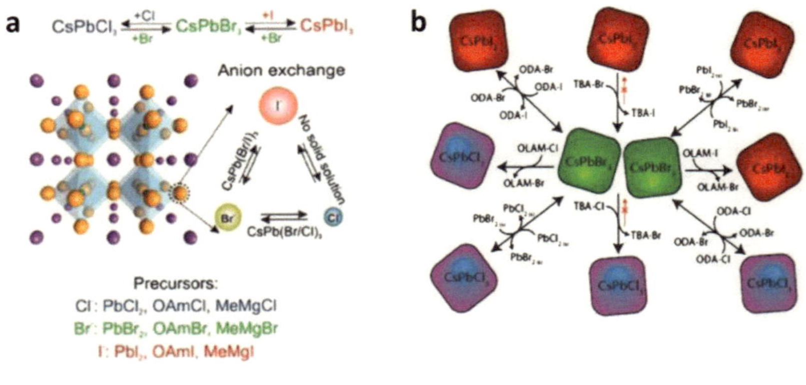

Figure 6. (a) Schematic representation of possible anion-exchange reactions within the crystal structure of $CsPbX_3$ crystal lattice, with indication of suitable reagents. (b) Different routes and precursors for $CsPbX_3$ (X = Cl, Br, I) ion exchange.

5. Optical properties of perovskite QDs

5.1. Exciton-related emission of perovskite QDs

Perovskite QDs exhibit remarkable optical properties, such as the high emission purity, large quantum yield, as well as tunable emission wavelength by varying the constituent elements, size, or dimensions. In the initial stage, most of the researches focused on the synthesis of the perovskite QDs and also their integration into device applications. But, it would be more interesting in our opinion if the optical properties of perovskite QDs or films made out of perovskite QDs could be investigated in detail. Despite the reported advances on perovskite-based optoelectronics, a deeper understanding of perovskites photophysical properties must be achieved if these materials are to make a technological impact. For example, as a key parameter controlling the recombination dynamics of photogenerated charges, the localization of exciton or free carrier needs to be specified. It means that the identification of dominant recombination mechanisms in perovskites will help interpret the seemingly counterintuitive facts that perovskites can act as both extraordinary photovoltaic materials and superior optical gain mediums for LED and lasers. In general, photovoltaic materials require efficient separation of photocarriers, and emissive materials require high recombination rates [28]. If the obtained exciton-binding energy of the perovskites is comparable to the thermal energy at RT (~26 meV), excited states will tend to dissociate into free carriers rather than recombination radiatively. The photoexcitation of perovskites can lead to the formation of free carriers and excitons, which then recombine to emit photons corresponding to the bandgap of perovskites. Although the question over the dominant species is still under investigation, it is generally accepted that free carriers are prevalent in perovskite films. Even et al. claimed that free carriers are photoexcited at RT, where Wannier-Mott excitons are dominant at low temperatures [29]. As for perovskite QDs, several groups have used temperature-dependent PL measurements to study the competition between exciton and free carriers. In Shi's study, temperature-dependent PL measurements were carried out to understand the optical transition mechanisms of the $CsPbBr_3$ QDs [30]. As shown in **Figure 7a**, with the decrease of the temperature, only one emission peak can be solved, indicating the absence of structural phase transition. And, the strong excitonic emission behavior was verified by performing the power-dependent PL measurements. As shown in **Figure 7b**, a power law dependence with $\beta = 1.31$ confirms the excitonic characteristics of the spontaneous emission. Moreover, by plotting the emission intensity of $CsPbBr_3$ QDs *versus* temperature, thermally activated nonradiative recombination process was observed, as shown in **Figure 7c**. Exciton-binding energy of the $CsPbBr_3$ QDs was further achieved by the following equation:

$$I(T) = \frac{I_0}{1 + A\exp\left(-\frac{E_B}{KT}\right)} \tag{1}$$

where K is the Boltzmann constant, T is the temperature, I_0 is the emission intensity at 0 K and A is a proportional constant. From the fitting, a value of 43.7 ± 4.9 meV for E_B was extracted. The data are much higher than the thermal energy at RT and ensure exciton survival well above RT. Therefore, they argued that the observed PL performance originates from the exciton-related emission. However, many studies attributed the PL spectra at RT to free carrier recombination. For example, D'Innocenzo et al. stated that excitons generated by low-density excitation were almost fully ionized at RT when the exciton-binding energy is moderately larger than the thermal energy at RT [31]. As shown in **Figure 7d**, the exciton-phonon interaction in carrier recombination process for $CsPbBr_3$ QDs was also investigated in Shi's report by studying the linewidth broadening behavior of $CsPbBr_3$ QDs. The value of optical phonon energy of 36.3 ± 1.8 meV was derived.

5.2. Quantum confinement effect in perovskite QDs

As we mentioned above, the optical properties of perovskite QDs depend not only on the constituent halide ions but also on their size or diameter. As we all know, quantum confinement effect has been widely studied in conventional semiconductor nanomaterials. Also, the

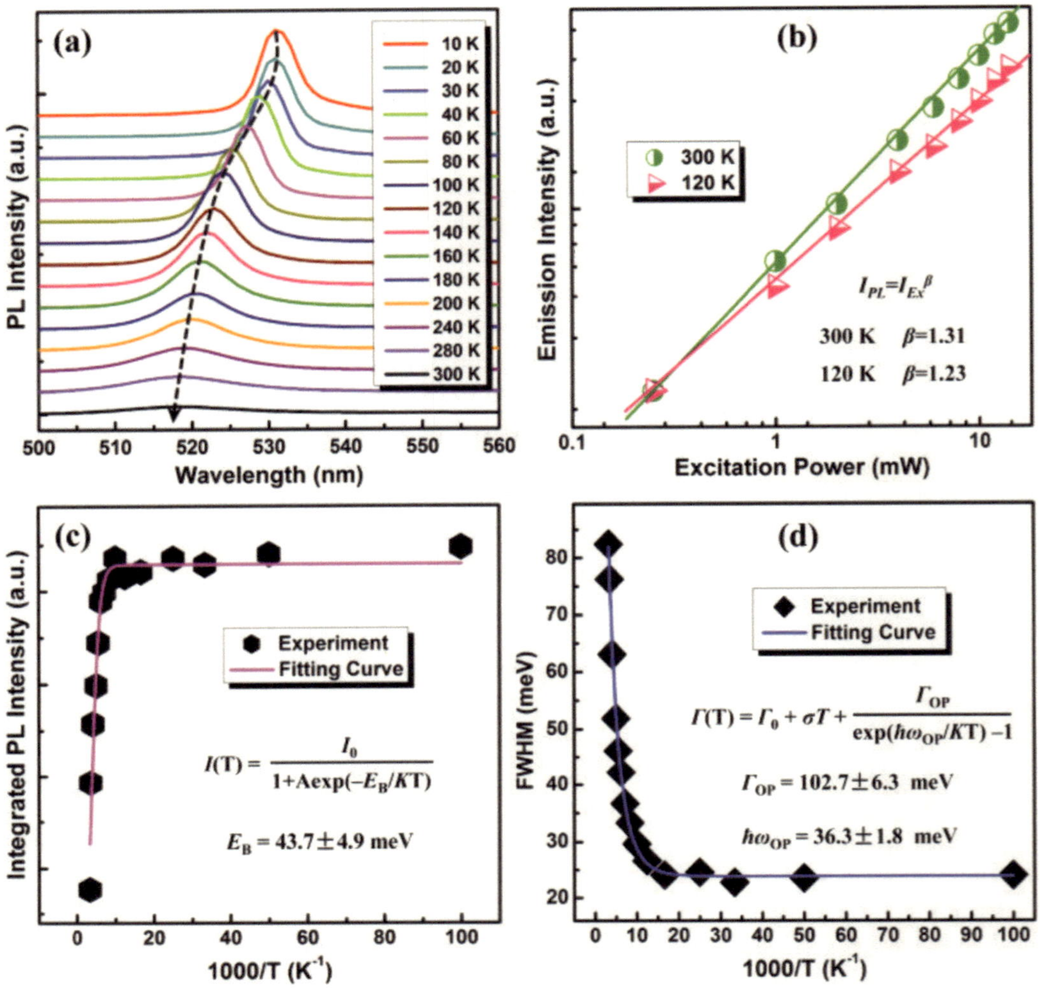

Figure 7. (a) Temperature-dependent PL spectra of the $CsPbBr_3$ QDs taken from 10 to 300 K. (b) The relationship between the integrated PL intensity and the excitation power of the $CsPbBr_3$ QDs at 300 and 120 K. (c) Integrated PL intensity and (d) linewidth of the $CsPbBr_3$ QDs as a function of reciprocal temperature from 10 to 300 K.

latest development in the size-controlled synthesis of perovskite QDs has enabled detailed investigations of quantum confinement effect in QDs. Friend and coworkers observed size-dependent photon emission from $MAPbBr_3$ QDs embedded in an organic matrix [16], where the QD size and their PL peak could be tuned by varying the concentration of the precursors. In their case, a $MAPbBr_3$ precursor solution (MABr and $PbBr_2$ dissolved in *N,N*-dimethylformamide) and 4,4-bis(*N*-carbazolyl)-1,1-biphenyl (CBP) matrix solution were prepared and mixed to achieve various weight ratios between CBP and the perovskite precursor. They firstly determined the average size of perovskite QDs by using X-ray diffraction, as shown in **Figure 8a**. As the concentration of perovskite QDs decreased, a reduced peak intensity accompanied by peak broadening was observed for (1 0 0) and (2 0 0) diffraction peaks, indicating a reduction of the size of nanocrystallites. Accordingly, the PL peak gradually shifted to higher energy decreasing the particle size, as shown in **Figure 8b**. They also summarized the changing trend by fitting with the equation of $E_{PL} = 2.39 + 12/d^2$ eV (**Figure 8c**), in which *d* is the particle size in nanometer. This suggested that the PL blueshift is a manifestation of quantum confinement of excitons in the perovskite nanocrystals. **Figure 8d** shows the corresponding absorbance data of the produced $MAPbBr_3$ QDs with different sizes, and a monotonic shift of the absorption edge toward the higher energies matched the above observation in PL spectra.

Quantum confinement effects have also been observed in all-inorganic $CsPbBr_3$ QDs. Kovalenko and coworkers reported the size-dependent PL emission from square-shaped $CsPbBr_3$ QDs [23], in which the PL emission peak gradually blueshifted from 512 to 460 nm as the diameter of QDs decreased from 11.8 to 3.8 nm, as shown in **Figure 9a**. **Figure 9b** shows the experimental and theoretical size dependence of the bandgap of $CsPbBr_3$ QDs in their case. In addition, tunable PL from $CsPbBr_3$ nanocrystals by varying the number of layers has also been reported, and the corresponding bandgap energy increases with the decreasing the number of layers. Alivisatos and coworkers reported the synthesis of quantum-confined highly fluorescent $CsPbBr_3$ nanoplatelets [32]. Their observations show that the thickness of $CsPbBr_3$ nanoplatelets can be tuned from 1 to 5 unit cells by changing the reaction temperature, with the monolayer platelets emitting at 400 nm, whereas the bulk-like crystals emitted at 520 nm.

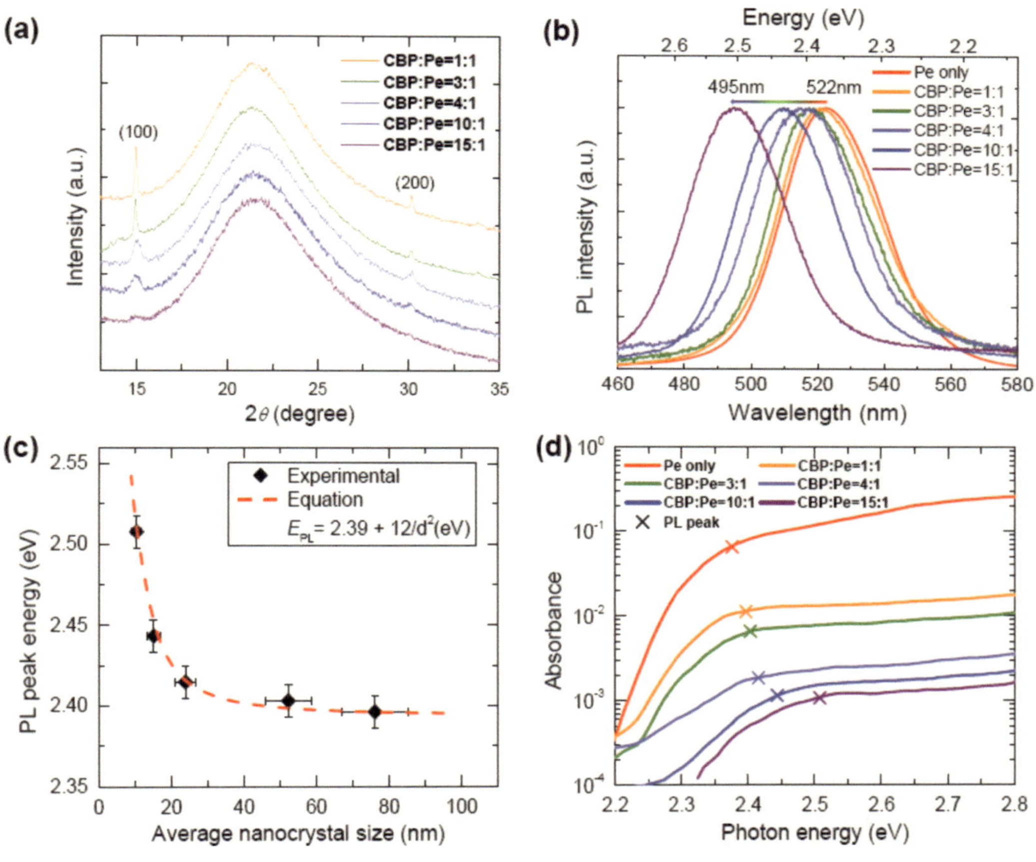

Figure 8. (a) X-ray diffraction patterns and (b) PL spectra of CBP:$MAPbBr_3$ with various weight ratios. (c) Energy of PL emission peak as a function of average perovskite nanocrystal size. (d) Absorbance data of the produced samples.

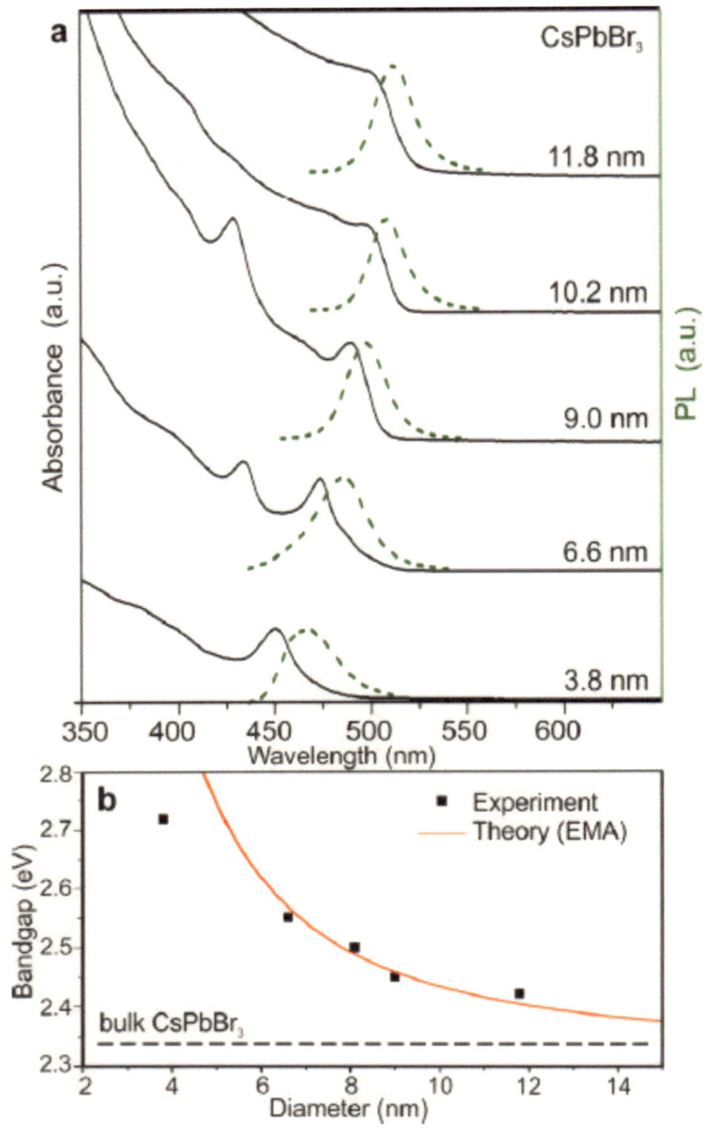

Figure 9. (a) Quantum size effects in the absorption and emission spectra of 5–12 nm $CsPbBr_3$ QDs. (b) Experimental *versus* theoretical (effective mass approximation (EMA)) size dependence of the bandgap energy.

6. Perovskite QD LEDs

6.1. Structure of perovskite QDs LEDs

As we mentioned above, perovskite QDs have been shown to possess high PL quantum yield, high emission purity, and tunable emission wavelength that make them suitable for high-performance, low-cost, and lightweight LEDs. Perovskite materials were incorporated into

http://dx.doi.org/10.5772/intechopen.68275

LEDs functioning at liquid nitrogen temperature back in the 1990s, and RT-working bright LEDs were fabricated by solution processing of hybrid perovskites by Tan et al. [33]. In their device, a solution-processed $MAPbI_{3-x}Cl_x$ perovskite layer was sandwiched between TiO_2 and poly(9,9-dioctylfluorene) (F8) layers for effective radiative recombination of electrons and holes in the perovskite layer. Also, using $MAPbBr_3$-emissive layer, an architecture of ITO/PEDOT:PSS/$MAPbBr_3$/F8/Ca/Ag was fabricated, producing a luminance of 364 cd/m^2 at a current density of 123 mA/cm^2. Because of the significantly higher quantum efficiency of perovskite QDs than that of perovskite bulk films, superior device performance for perovskite QD LEDs could be expected [34]. A typical perovskite QD LED consists of an intrinsic active layer in a double-heterojunction structure. As shown in **Figure 10**, an n-type electron transport layer and a p-type hole transport layer are usually used to construct a typical perovskite QD LED. Under forward bias, the injection of charge carriers into a thin luminescent layer leads to radiative recombination and provides light emission in all directions. Efficient LEDs use electrodes that readily inject carriers into the active region and prevent charges from passing through the device and quenching at contacts. Since the work of Tan et al. in 2014 [33], various perovskite QD LEDs using different perovskite active layers and electron/hole transport layers have been reported.

6.2. LEDs based on perovskite QDs

Song and coworkers firstly reported the perovskite QD LEDs with tunable emission wavelength based on $CsPbBr_3$ QDs [4]. For the device structure of ITO/PEDOT:PSS/PVK/QDs/TPBi/LiF/Al (shown in **Figure 11a**), the luminescence of blue, green, and orange LEDs reached 742, 946, and 528 cd/m^2, with external quantum efficiency of 0.07, 0.12, and 0.09%, respectively. In addition, the produced LEDs possess narrow linewidths, indicating their potential applications in displays. However, non-optimized devices exhibited poor emission efficiency. Generally, in order to improve the device performance, three important factors should be considered: carrier

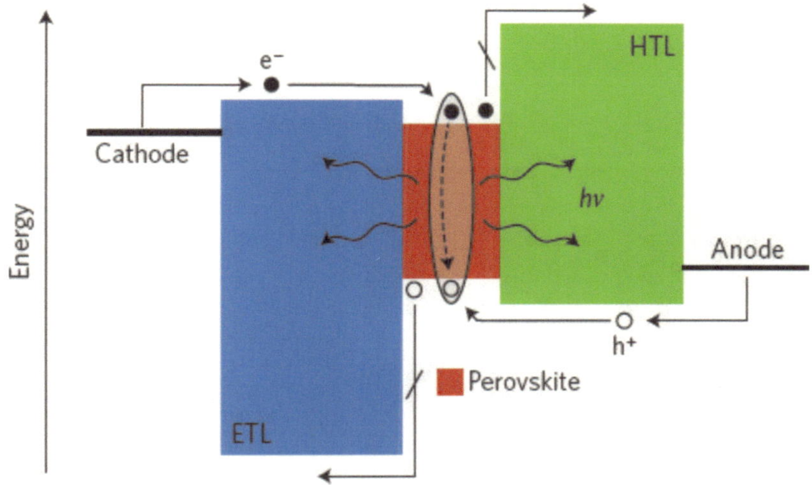

Figure 10. General device structure of perovskite QD LEDs. ETL, electron transport layer; HTL, hole transport layer.

injection efficiency, radiative recombination efficiency, and injection balance. For an improved carrier injection efficiency, a thin perfluorinated ionomer film was introduced between the hole transport layer and $CsPbBr_3$ QDs active layer in Zhang's study [35]. In their case, the hole injection efficiency was enhanced greatly, which favored a high brightness. Besides, the usage of carrier transport layers with a high conductivity could ensure a high carrier injection efficiency, resulting in an improved external quantum efficiency.

In order to promote the carrier radiative recombination efficiency, the undesired lattice and surface/interface defects should be excluded. Sun's research group found an interesting phenomenon that the introduction of $CsPb_2Br_5$ QDs attached on $CsPbBr_3$ QDs could improve the emission lifetime by decreasing nonradiative energy transfer to the trap states via controlling the trap density [36]. As a result, a high external quantum efficiency of about 2.21% was achieved, and a maximum luminance of 3853 cd/m^2 was obtained. **Figure 12a** shows the X-ray diffraction patterns of the synthesized perovskite QD products. The peak located at 30.36° is assigned to (2 0 2) diffraction of $CsPbBr_3$, whereas the peak at 30.69° is identified from $CsPb_2Br_5$. And, the impurities of $CsPb_2Br_5$ phase were emerged in the low-temperature (70°C) solution-processed $CsPbBr_3$ products. The generation of the secondary phase $CsPb_2Br_5$ in the product can be ascribed to the following process: $PbBr_2 + CsPbBr_3 \rightarrow CsPb_2Br_5$. **Figure 12b** displays the schematic diagrams of the corresponding crystal structures of the $CsPbBr_3$ and $CsPb_2Br_5$ QDs. They further fabricated the LED device by constructing the heterostructure shown in

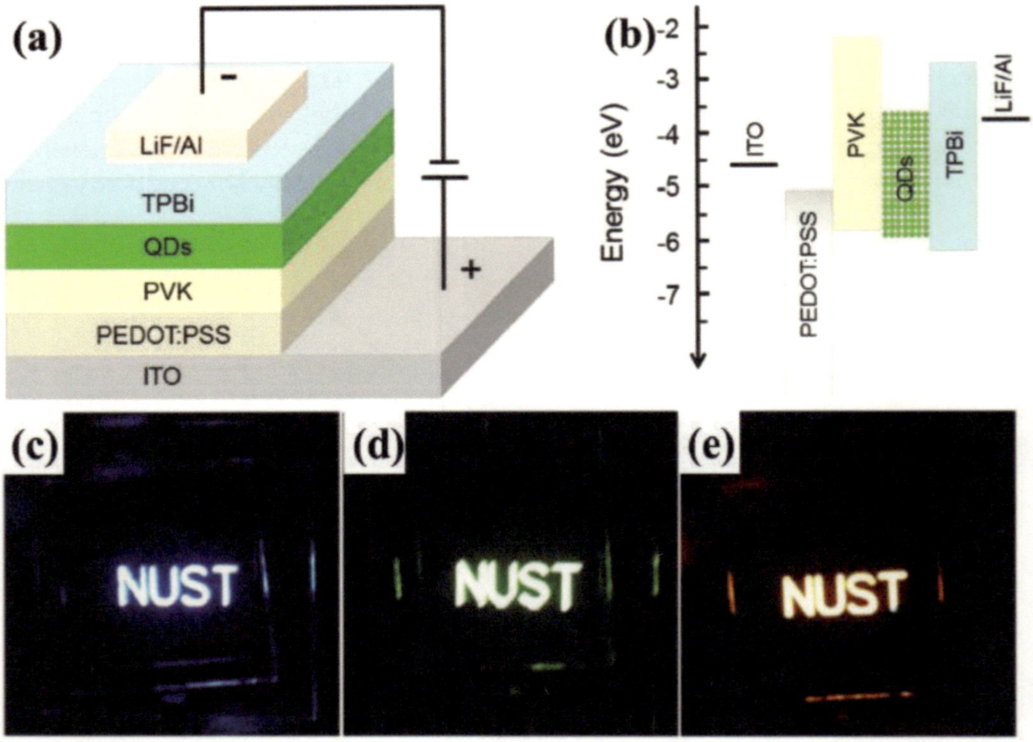

Figure 11. (a) Structure of the perovskite LED device. (b) Energy band alignment of the device. (c–e) Blue, green, and orange LEDs with abbreviations of NUST.

Figure 12c and achieved a high-purity green emission at about 527 nm. The inset of **Figure 12d** shows the corresponding photograph of a device with an active area of 2 × 2 mm². In fact, contradiction exists about carrier radiative recombination efficiency and injection efficiency. For example, to keep a high quantum yield, surface passivation is usually necessary, while long-chain ligands will reduce the conductivity of QDs. In this regard, Pan and coworkers applied a short ligand, di-dodecyl dimethyl ammonium bromide, to passivate $CsPbBr_3$ QDs and facilitate the carrier transport ability [37]. Consequently, a promoted PL quantum yield of about 71% was achieved, and a higher external quantum efficiency of ~3.0% was obtained.

As discussed above, the carrier injection efficiency, radiative recombination efficiency, and injection balance are important factors for the device performances of perovskite QD LEDs. In fact, the QDs applied in previous reports were not well purified due to the continuous growth with classical purification strategies. Recently, Zeng's group developed a handy surface purification method that can achieve recyclable treatment on QDs using hexane-/ethyl acetate-mixed solvent [38]. Thus, a balance of carrier injection efficiency and surface passivation effect can be constructed, as illustrated in **Figure 13a**. After two purifications, the QDs

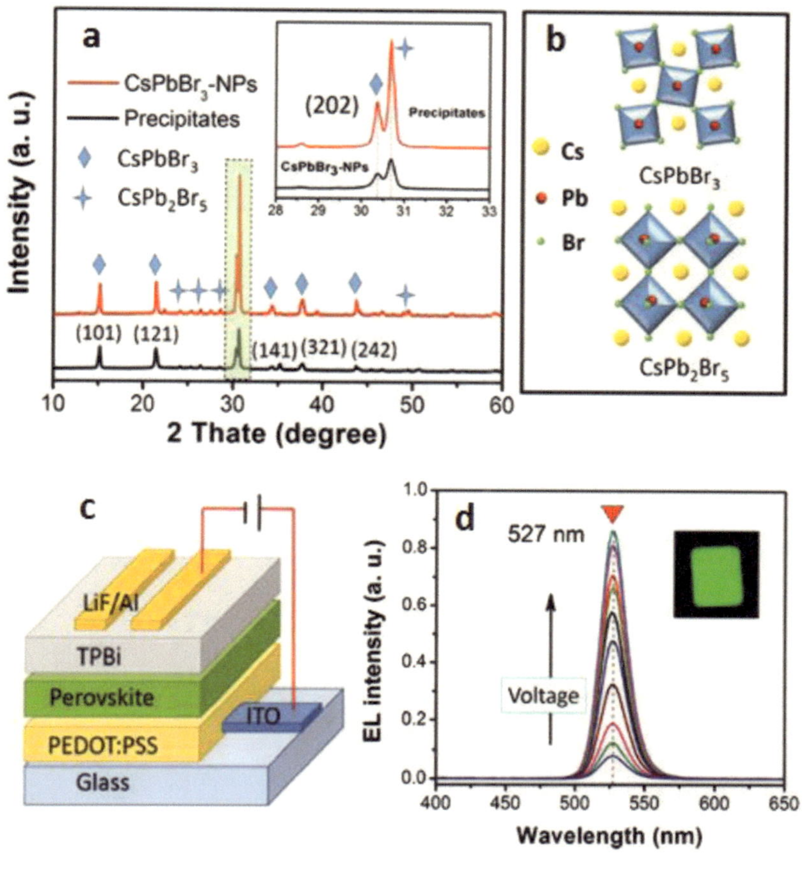

Figure 12. (a) X-ray diffraction patterns of $CsPbBr_3$ QDs and precipitates of products. (b) The schematic diagrams of the corresponding crystal structures of the $CsPbBr_3$ and $CsPb_2Br_5$ QDs. (c) Device structure of the studied perovskite LEDs. (d) EL spectra at different driving voltages. The inset shows the photograph of a device with an active area of 2 × 2 mm².

obtained still possess a good stability (shown in **Figure 13b**). But, the PL quantum yield shows no obvious decrease (**Figure 13c**), indicating a good surface passivation. Therefore, we could expect that good dispersivity ensures an excellent film formation ability and pure surface that leads to efficient carrier transport. So, the efficiency of carrier injection was greatly enhanced (shown in **Figure 13d**). In virtue of the above advantages discussed, the luminance (**Figure 13e**) and current efficiency (**Figure 13f**) of the resulting LEDs based on $CsPbBr_3$ QDs are greatly improved compared to previous studies, and more importantly, the studied LED achieves a record external quantum efficiency of 6.27% (**Figure 13g**).

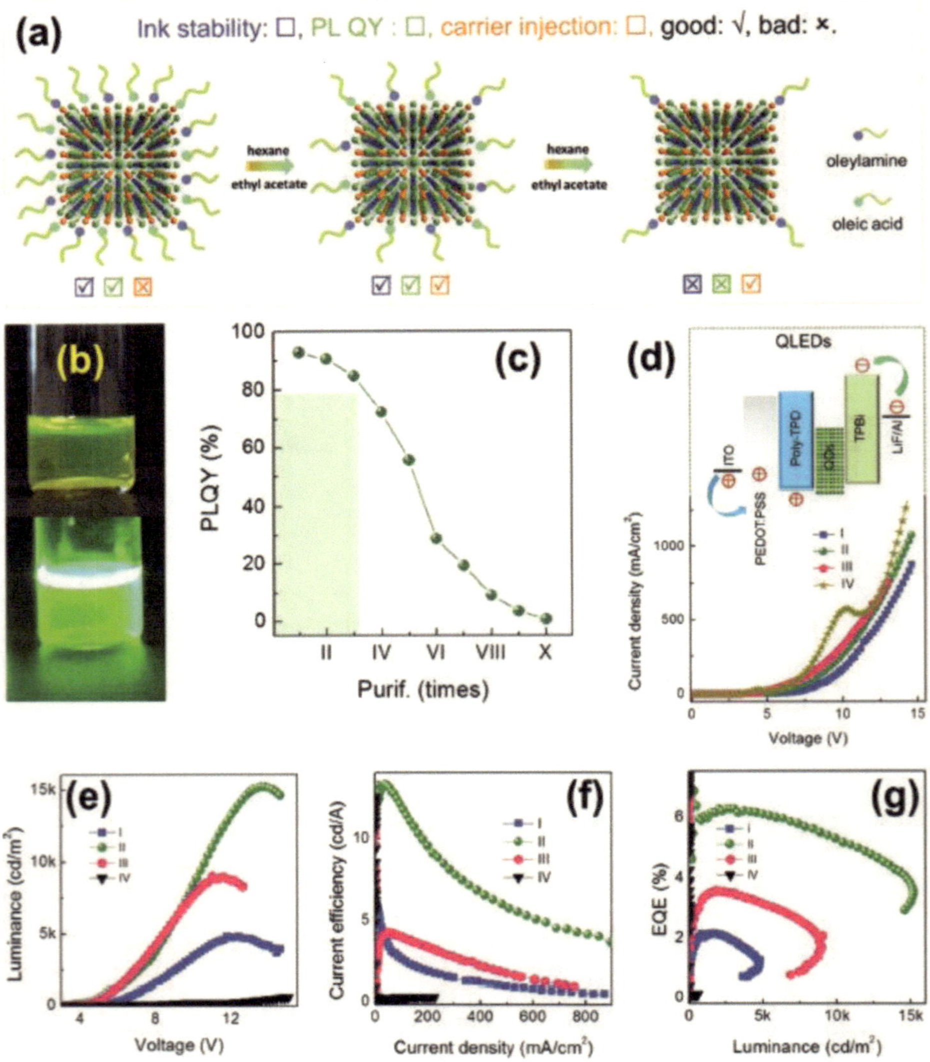

Figure 13. (a) Schematic illustration of the control of ligand density of $CsPbBr_3$ QD surfaces. (b) Photographs of $CsPbBr_3$ QD inks without (upper) and with (below) UV light excitation after two cycles of treatments. (c) PL quantum yield of $CsPbBr_3$ QDs with different purifying cycles. (d) Current density-voltage, (e) luminance-voltage, (f) current efficiency-current density, and (g) external quantum efficiency-luminance characteristics of the perovskite QD LEDs.

Although the device performance of perovskite QD LEDs was promoted greatly in a short time, operation stability has always been criticized for such devices in a continuous current mode, which are the main obstacles hindering the reliable device operation and their future application. In previous reports, conventional carrier injection conducting polymer or small molecules, such as PEDOT:PSS, PCBM, and 1,3,5-tris(2-*N*-phenylbenzimidazolyl) benzene, have been frequently employed as the carrier injectors in perovskite LEDs [25, 33, 35, 36], but their inherent chemical instability inevitably degrades the device performance; thus, a high-efficiency light emission cannot be sustained over a long running time. More recently, Shi et al. present a strategy that addresses simultaneously the emission efficiency and stability issues facing current perovskite LEDs' compromise [30]. Wide bandgap semiconductors, n-MgZnO and p-MgNiO, were employed as the electron and hole injectors to construct $CsPbBr_3$ QD LEDs. The resulting diode demonstrates a high luminance (3809 cd/m^2) and external quantum efficiency (2.39%), as well as a significantly improved stability compared with reference and other previously reported devices constructed with organic carrier injectors. **Figure 14a** illustrates the emission intensity of the diode *versus* running time, and one can see that the EL intensity has almost not changed over 30 min. A long-term operation stability measurement demonstrated that the device could operate continuously for 10 h with an emission decay of below 20%, greatly superior to other previously reported devices constructed with organic electron and/or hole transport layers. It is believed that the device concept proposed in their study will provide valuable information for the future design and development of high-efficiency and air-stable perovskite QD LEDs.

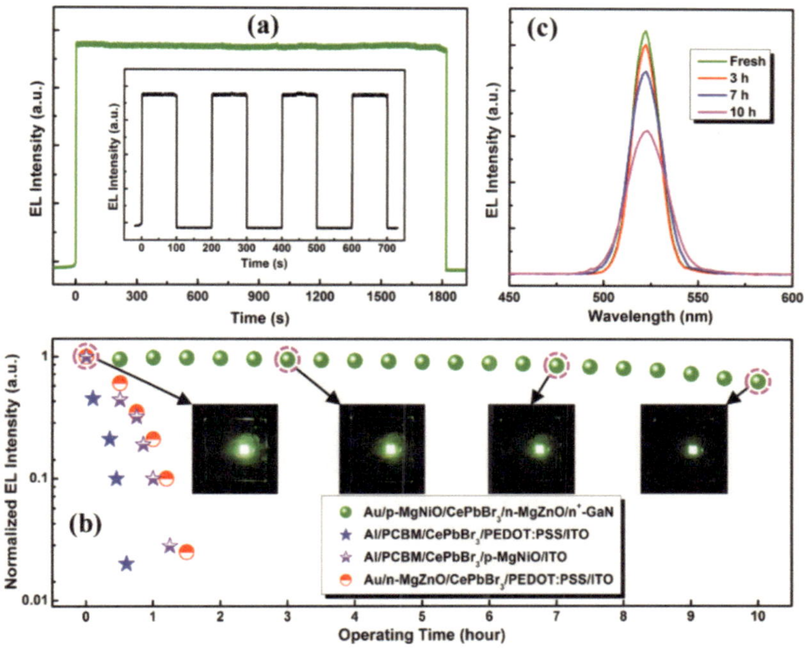

Figure 14. (a) Time dependence (30 min) of the emission intensity of the prepared LEDs at 8.0 V. (b) Emission intensity of the studied LEDs and three reference LEDs as a function of running time under a bias of 10.0 V. The insets show the corresponding photographs of the LEDs after different running periods. (c) Evolution of the emission spectra of the prepared LEDs after different running periods.

7. Summary

In summary, recent developments in perovskite QDs add a new class of members to the family of colloidal QDs. In spite of a short development of only 2 years, perovskite QDs have shown great potential in application in cost-effective LED fields. However, compared with traditional metal chalcogenide QDs, novel synthesis approaches are needed to tailor the surface ligands and enhance the stability of perovskite QDs. At present, the perovskite LEDs based on both organic-inorganic hybrid and inorganic QDs show relatively poor performances, and, therefore, new device structures are required to make a step change in the emission efficiency and operation stability of LEDs. Despite these challenges, we believe that the newly emerging perovskite QDs have a bright future, and the previous studies will provide valuable information for the future design and development of high-performance perovskite QD LEDs.

Author details

Zhifeng Shi*, Xinjian Li and Chongxin Shan

*Address all correspondence to: shizf@zzu.edu.cn

Department of Physics and Engineering, Zhengzhou University, Zhengzhou, China

References

[1] Coe S, Woo WK, Bawendi M, Bulovic V. Electroluminescence from single monolayers of nanocrystals in molecular organic device. Nature. 2002;**420**:800-803

[2] Colvin VL, Schlamp MC, Alivsistors AP. Light-emitting diodes made from cadmium selenide nanocrystals and a semiconducting polymer. Nature. 1994;**370**:354-357

[3] Shirasaki Y, Supran GJ, Bawendi MG, Bulovic V. Emergence of colloidal quantum-dot light-emitting technologies. Nature Photonics. 2013;**7**:13-23

[4] Song J, Li J, Li X, Xu L, Dong Y, Zeng H. Quantum dot light-emitting diodes based on inorganic perovskite cesium lead halides ($CsPbX_3$). Advanced Materials. 2015;**27**: 7162-7167

[5] Qian L, Zheng Y, Xue J, Holloway PH. Stable and efficient quantum-dot light-emitting diodes based on solution-processed multilayer structures. Nature Photonics. 2011;**5**:543-548

[6] Lee KH, Lee JH, Song WS, Ko H, Lee C, Lee JH, Yang H. High efficiency, color-pure, color-stable blue quantum dot light-emitting devices. ACS Nano. 2013;**7**:7295-7302

[7] Shen H, Cao W, Shewnon NT, Yang CC, Li LS, Xue JG. High-efficiency, low turn-on voltage blue-violet quantum-dot-based light-emitting diodes. Nano Letters. 2015;**15**:1211-1216

[8] Jin SY, Lian TQ. Electron transfer dynamics from single CdSe/ZnS quantum dots to TiO_2 nanoparticle. Nano Letters. 2009;**9**:2448-2454

[9] Vybornyi O, Yakunin S, Kovalenko MV. Polar-solvent-free colloidal synthesis of highly luminescent alkylammonium lead halide perovskite nanocrystals. Nanoscale. 2016;**8**: 6278-6283

[10] Buim A, Comin R, Ip AH, Sargent EH. Perovskite quantum dots modeled using ab initio and replica exchange molecular dynamics. Journal of Physical Chemistry C. 2015;**119**: 13965-13971

[11] Tyagi P, Arveson SM, Tisdale WA. Colloidal organohalide perovskite nanoplatelets exhibiting quantum confinement. Journal of Physical Chemistry Letters. 2015;**6**:1911-1916

[12] Wang Y, Li XM, Zhao X, Xiao L, Zeng H, Sun H. Nonlinear absorption and low-threshold multiphoton pumped stimulated emission from all-inorganic perovskite nanocrystals. Nano Letters. 2016;**16**:448-453

[13] Huo C, Cai B, Yuan Z, Ma BW, Zeng HB. Two-dimensional metal halide perovskites: Theory, synthesis, and optoelectronics. Small Methods. 2017;1:1600018

[14] Kulbak M, Gupta S, Levine I, Hodes G, Cahen D. Cesium enhances long-term stability of lead bromide perovskite based solar cells. Journal of Physical Chemistry Letters. 2016;**7**:167-172

[15] Huang H, Polavarapu L, Sichert JA, Susha A, Urban A, Rogach A. Colloidal lead halide perovskite nanocrystals: Synthesis, optical properties and applications. NPG Asia Materials. 2016;**8**:e328

[16] Di D, Musselman K, Li G, Sadhanala A, Levskaya Y, Song Q, Tan Z, Lai M, Driscoll J, Greenham N, Friend RH. Size-dependent photon emission from organometal halide perovskite nanocrystals embedded in an organic matrix. Nano Letters. 2015;**6**:446-450

[17] Schmidt L, Pertegas A, Carrero S, Malinkiewicz O, Agouram S, Espallargas M, Bolink H, Galian R, Perez-Prieto J. Nontemplate synthesis of $CH_3NH_3PbBr_3$ perovskite nanoparticles. Journal of the American Chemical Society. 2014;**136**:850-853

[18] Gonzalez S, Galian R, Perez-Prieto J. Maximum the emissive properties of $CH_3NH_3PbBr_3$ perovskite nanoparticles. Journal of Materials Chemistry A. 2015;**3**:9187-9193

[19] Li X, Wu Y, Zhang S, Cai B, Gu Y, Song J, Zeng H. $CsPbX_3$ quantum dots for lighting and displays: Room-temperature synthesis, photoluminescence superiorities, underlying origins and white light-emitting diodes. Advanced Functional Materials. 2016;**26**:2435-2445

[20] Huang H, Zhao F, Liu L, Zhang F, Wu X, Shi L, Zou B, Pei Q, Zhong H. Emulsion synthesis of size-tunable $CH_3NH_3PbBr_3$ quantum dots: An alternative route toward efficient light-emitting diodes. ACS Applied Materials and Interfaces. 2015;**7**:28128-28133

[21] Luo B, Pu Y, Yang Y, Lindley S, Abdelmageed G, Ashry H, Li Y, Li X, Zhang J. Synthesis, optical properties, and exciton dynamics of organolead bromide perovskite nanocrystals. Journal of Physical Chemistry C. 2015;**119**:26672-26682

[22] Zhang F, Zhong H, Chen C, Wu X, Hu X, Huang H, Han J, Zou B, Dong Y. Brightly luminescent and color-tunable colloidal $CH_3NH_3PbBr_3$ (X = Br, I, Cl) quantum dots: Potential alternatives for display technology. ACS Nano. 2015;**9**:4533-4542

[23] Protesescu L, Yakunin S, Bodnarchuk M, Krieg F, Caputo R, Hendon C, Yang R, Walsh A, Kovalenko MV. Nanocrystals of cesium lead halide perovskites ($CsPbBr_3$, X = Cl, Br, and I): Novel optoelectronic materials showing bright emission with wide color gamut. Nano Letters. 2015;**15**:3692-3696

[24] Jang D, Park K, Kim D, Park J, Shaojaei F, Kang H, Ahn J, Lee J, Song J. Reversible halide exchange reaction of organometal trihalide perovskite colloidal nanocrystals for full-range band gap tuning. Nano Letters. 2015;**15**:5191-5199

[25] Wong A, Lai M, Eaton S, Yu Y, Lin E, Dou L, Fu A, Yang P. Growth and anion exchange conversion of $CH_3NH_3PbX_3$ nanorod arrays for light-emitting diodes. Nano Letters. 2015;**15**:5519-5524

[26] Nedelcu G, Protesescu L, Yakunin S, Bodnarchuk M, Grotevent M, Kovalenko M. Fast anion-exchange in highly luminescent nanocrystals of cesium lead halide perovskites ($CsPbX_3$, X = Cl, Br, I). Nano Letters. 2015;**15**:5635-5640

[27] Akkerman Q, D'Innocenzo V, Accornero S, Scarpellini A, Petrozza A, Prato M, Manna L. Tuning the optical properties of cesium lead halide perovskite nanocrystals by anion exchange reactions. Journal of the American Chemical Society. 2015;**137**:10276-10281

[28] He H, Yu Q, Li H, Li J, Si J, Jin Y, Wang N, Wang J, He J, Wang X, Zhang Y, Ye Z. Exciton localization in solution-processed organolead trihalide perovskites. Nature Communication. 2015;**7**:10896

[29] Even J, Pedesseau L, Katan C. Analysis of multivalley and multibandgap absorption and enhancement of free carriers related to exciton screening in hybrid perovskites. Journal of Physical Chemistry C. 2014;**118**:11566-11572

[30] Shi Z, Li Y, Zhang Y, Chen Y, Li X, Wu D, Xu T, Shan C, Du G. High-efficiency and air-stable perovskite quantum dots light-emitting diodes with an all-inorganic heterostructure. Nano Letters. 2017;**17**:313-321

[31] D'Innocenzo V, Grancini G, Alcocer M, Kandada A, Stranks S, Lee M, Lanzani G, Snaith H, Petrozza A. Exciton *versus* free charges in organo-lead tri-halide perovskites. Nature Communication. 2014;**5**:3586-3591.

[32] Bekenstein Y, Koscher B, Eaton S, Yang P, Alivisatos A. Highly luminescent colloidal nanoplates of perovskite cesium lead halide and their oriented assemblies. Journal of the American Chemical Society. 2015;**137**:16008-16011

[33] Tan Z, Moghaddam R, Lai M, Docampo P, Higler R, Deschler F, Price M, Sadhanala A, Pazos L, Credgington D, Hanusch F, Bein T, Snaith H, Friend R. Bright light-emitting diodes based on organometal halide perovskite. Nature Nanotechnology. 2014;**9**:687-692

[34] Sutherland B, Sargent E. Perovskite photonic sources. Nature Photonics. 2016;**10**:295-302

[35] Zhang X, Lin H, Huang H, Reckmeier C, Zhang Y, Choy W, Rogach A. Enhancing the brightness of cesium lead halide perovskite nanocrystal based green Light-Emitting Devices through the interface engineering with perfluorinated ionomer. Nano Letters. 2016;**16**:1415-1420

[36] Zhang X, Xu B, Zhang JB, Gao Y, Zheng Y, Wang K, Sun X. All-inorganic perovskite nanocrystals for high-efficiency light-emitting diodes: Dual-phase $CsPbBr_3$-$CsPb_2Br_5$ composites. Advanced Functional Materials. 2016;**26**:4595-4600

[37] Pan J, Sarmah S, Murali B, Dursun I, Peng W, Parida M, Liu J, Sinatra L, Alyami N, Zhao C, Alarousu E, Ng TK, Ooi B, Bakr O, Mohammed O. Air-stable surface-passivated perovskite quantum dots for ultra-robust, single- and two-photon-induced amplified spontaneous emission. Journal of Physical Chemistry Letters. 2015;**6**:5027-5033

[38] Li J, Xu L, Wang T, Song J, Chen J, Xue J, Dong Y, Cai B, Shan Q, Han B, Zeng H. 50-fold EQE improvement up to 6.27% of solution-processed all-inorganic perovskite $CsPbBr_3$ QLEDs via surface ligand density control. Advanced Materials. 2017;**29**:1603885

Chapter 5

CdS(Se) and PbS(Se) Quantum Dots with High Room Temperature Quantum Efficiency in the Fluorine-Phosphate Glasses

Elena Kolobkova and Nikolay Nikonorov

Additional information is available at the end of the chapter

http://dx.doi.org/10.5772/intechopen.68459

Abstract

In this study, the luminescent properties of the CdS(Se) quantum dots (QDs) with the mean size of 2–4 nm, the (PbS)n/(PbSe)n molecular clusters (MCs) and PbS(Se) quantum dots (QDs) with the mean size of 2–5 nm embedded in the fluorine-phosphate glass are investigated. The dependence of the photo luminescence absolute quantum yield (PL AQY) on the sizes of the CdS(Se) QDs are studied. It is found that the PL AQY of the CdSe QDs increases monotonically to its maximum height and then falls down. The PL AQY of the CdS(Se) QDs can reach 50–65%. Luminescence of (PbS)n/(PbSe)n MCs embedded in fluorine-phosphate matrix and excited by UV radiation is obtained in the visible spectral region and its absolute quantum yield was up to 10%. The PbSe QDs have broadband photoluminescence with quantum efficiency about five times more than MCs (~50%) in the spectral range of 1–1.7 μm. Glasses doped with PbS(Se) QDs provide potential application as infrared fluorophores, which are both efficient and possess short life times. The glass matrix protects the QDs from external influence and their optical properties remain unchanged for a long time.

Keywords: quantum dots, molecular clusters, fluorine-phosphate glass, luminescence, absolute quantum yield

1. Introduction

1.1. Why fluorine-phosphate glasses?

Glass-crystalline materials (or glassceramics) are composite materials, usually consisting of a glassy matrix (glass phase) and nano-sized dielectric or semiconductor crystals or metallic particles distributed in it. Glassceramics are formed by the growth of crystalline phase inside of a glass matrix. The growth of crystalline phase occurs as a result of the crystallization of amorphous phase during heat treatment. Glasses doped with CdS, CdSe, CdTe, PbS, PbSe nanocrystals are typical representatives of the nano-glassceramics or nanocomposite materials and usually used as a nonlinear optical media [1, 2]. Traditionally silicate glassy matrix has been used for QDs formation. These glasses are characterized by high synthesis temperature (1300–1500°C), long-time (tens of hours) and high temperature (500–700°C) of heat treatment [3–7].

In this study, we used a fluorine-phosphate (FP) glass featuring with a number of advantages compared to conventional silicate glasses, including low temperature synthesis and heat treatment, formation of quantum dots in the wide range of concentrations, short time of heat treatment and higher spatial distribution homogeneity. Two-stage heat treatment was afforded to get quantum dots (QDs) with narrow size distribution. The CdS and CdSe QDs dispersed in a fluorine-phosphate glasses are attracting much attention as nonlinear optical materials [1], but information about luminescent properties is not available.

In this communication, we present the results of our research designed for synthesis and characterization of yellow-red and IR emitting semiconductor quantum dots embedded in the fluorine-phosphate glass. The composites are fabricated by two-staged heat treatment and stabilization of synthesized II–VI and IV–VI semiconductor QDs into glass matrices. **Figure 1** demonstrates initial glasses and glasses after heat treatment.

1.2. Why phosphors for visible spectral region based on CdSe and CdS quantum dots?

QDs are a type of nanomaterials with good fluorescent properties. The size-dependent emission is probably the most attractive property of semiconductor nanocrystals. Among them, CdS and CdSe QDs are one of the most promising materials because QDs have bright luminescence in the visible range of the optical spectrum. For example, CdSe QDs have shown potential as superior biological labels [8–10], laser sources [11, 12] and tunnel diodes [13, 14].

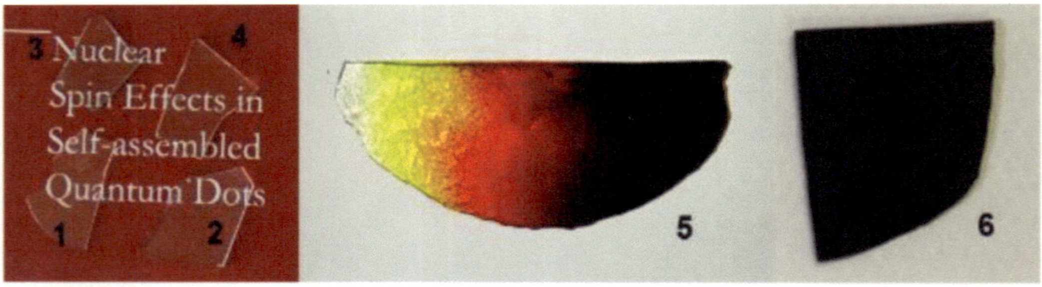

Figure 1. Photo Initial glasses doped with CdS (1), CdSe (2), PbS (3) and PbSe (4) and glasses after heat treatment doped with CdSe QDs (with different sizes) (5) and PbSe (6).

Compared with conventional fluorescent dyes, CdS(Se) QDs have a wide continuously distributed excitation spectra, not only with symmetrical distributed narrow emission spectra, but also many other excellent properties such as adjustable color, excellent photochemical stability and high threshold of light bleaching [15].

However, colloidal synthesized bare quantum dots, including CdSe(S) QDs, usually have surface defects and cause to reduction in absolute photoluminescence (PL) quantum yield. The best PL absolute quantum yield (AQY) (quantum efficiency) reported for the as-prepared nanocrystals at room temperature is around 20% in the wavelength range of 520–600 nm and about a few percent or lower in the spectrum range above 600 nm. Also, in the spectrum range below 520 nm, it is a few percent or lower of total absolute quantum yield [13, 15]. In general, a low PL AQY is due to the surface states located in the bandgap of the nanocrystals. These surface states act as trapping states for the photogenerated charges. The origin of the surface trapping states is dangling bonds of the surface atoms [3, 16, 17]. Therefore, it is essential to control the QDs surfaces to reduce the surface defects by passivating the surface of QDs [18, 19].

The core/shell structures solve optical problems, such as low PL AQY and improve the stability of QDs. As was shown, the room temperature quantum efficiency of the band-edge luminescence of CdSe QDs could be improved to 40–60% by surface passivation with inorganic (ZnS) or organic (alkylamines) shells [18]. Due to the non-radiative processes, a reduction in the PL emission intensity was observed by annealing of the prepared samples in different environments (oxygen, hydrogen, and air).

Due to the failure of orange-red emitting materials in general, efforts were directed to synthesize colloidal CdS/CdSe QDs that can emit light with wavelengths that range from orange (600 nm) to red (650 nm). Nevertheless, both stability and reproducibility of the PL AQY are not predictable. With some inorganic and organic surface functionalization, the PL QY of synthesized colloidal CdSe nanocrystals is boosted more than 50% in the range of 520–650 nm. However, the emission efficiency for the orange-red color window is still low. Especially for red emission (around 650 nm), the PL QY of the nanocrystals in solution was nearly zero [15].

1.3. Why optical transparent glasses doped with CdS and CdSe quantum dots?

Semiconductor CdS and CdSe nanoparticles dispersed in a silicate glass matrix are attracting much attention [3–5]. The possibility of QDs formation in the optical glasses creates significant benefits for their application. Currently, optical transparent glasses doped with nanocrystals are of great interest for the modern element base of photonics. In these materials, the best properties of nanocrystals and glasses technology (possibilities of pressing and molding, spattering, pulling optical fibers, using ion exchange technology) are easily combined. In addition, the glass matrix protects the QDs from external influence.

In the QDs doped silicate glasses, the PL spectra consist of two bands: a less intense high-energy band and a lower energy broader band. The first band occurs at a wavelength 10–20 nm higher than the absorption edge of QDs and is due to direct electron-hole recombination. This weak peak shifts to the higher wavelength with increasing particle size. The second band is due to surface defects and occurs at 800–900 nm spectral region. PL AQY is less than 30% for CdS(Se) QDs in the silicate glasses and decreases as the size of QD increases [4].

1.4. Optical composite materials with PbS and PbSe quantum dots as tunable near-IR phosphors

Semiconductor PbS and PbSe quantum dots have been widely investigated because of their unique optical and electronic properties. PbS and PbSe are a typical narrow bandgap ($E_g = 0.41$ eV, $E_g = 0.29$ eV) semiconductors with the Bohr's exciton radii 20 nm (for PbS) and 46 nm (for PbSe). Their large Bohr radius provides strong quantum confinement effect that can be easily realized in a wide range of QD size [20, 21].

Optical composite materials with PbS and PbSe QD are widely used in optical and photonic applications. For example, PbS(Se) QDs have potential applications in near-infrared photodetectors, photovoltaics and as saturable absorbers [22, 23]. In contrast to rare-earth ions-doped optical materials, lead chalcogenide quantum dot-doped glasses can provide tunable near-IR luminescence (about 1–2 μm) with high absolute quantum yield (≥60%) by controlling the size and distribution of QDs [22–24]. Despite a lot of efforts spent to find a novel gain medium which can almost cover the whole optical communication window, it is essential to tailor the emission of QDs embedded in a glass matrix in a wide bandwidth of the optical communication network. Therefore, PbS and PbSe QDs are promising gain media for covering the whole optical communication window [24–28]. Several advances of controlled chemical synthesis of the materials have provided the ways to grow QDs and manipulate their size, shape and composition using different methodologies [29].

PbSe and PbS QDs with a band gap in the mid-infrared are interesting infrared phosphors materials. In the visible spectral range organic dyes usually present large PL AQY value than quantum dots, but infrared organic dyes have very low PL AQY suffer from poor photostability. On the other hand, the rare-earth ions can be used as the efficient infrared phosphors but they have small absorption cross sections and microsecond lifetimes. Infrared phosphors of PbS(Se), that are both efficient and possess short lifetimes, may find unique applications in the near-IR spectral range where biological tissues are relatively transparent, or as fluorescent materials in the fiber communication range of 1.3–1.5 microns.

Variation of temperature and duration heat treatment (HT) of the glasses doped with Pb, S or Se make possible to change sizes of PbS(Se) QDs in a wide range (2–15 nm). The choice of these parameters enhances the formation of nucleation centers (NCs) in the glass and leads to good size control and small size dispersions. The growth processes of lead chalcogenide QDs is based mainly on the phase decomposition from the over-saturated solid solution. Two-stage heat treatment of the glass afforded QDs with narrow size distribution.

Lead chalcogenide QDs have been synthesized in silicate glasses, fluorine-phosphate glasses and germanosilicate glasses [7, 30–32]. Among these glasses, fluorine-phosphate glasses have the largest solubility of lead chalcogenide compounds [32].

It is evident that during the process of formation and growth, these structures pass through the phase when their sizes are less than 1.0 nm and they do not possess semiconductor properties. This phase corresponds to the existence of lead chalcogenide compounds in a form of molecular clusters (MCs) [36]. The optical properties of MCs are considerably different from the properties of QDs.

In this study, we used a fluorine-phosphate glass (FPG) doped with high concentration of CdS, CdSe, PbSe and PbS QDs. FPGs are featured with a number of advantages compared to conventional silicate and borosilicate glasses. FPGs characterize low temperature synthesis, possibility for a wide range adjustment of the formed quantum dots concentration, low temperature and time of heat treatment and higher spatial distribution homogeneity of the QDs. We represent the luminescent properties of the fluorine-phosphate glasses doped with CdS, CdSe, PbSe and PbS QDs.

2. Preparation procedure

In order to investigate the effect of the QDs sizes on the PL properties, the fluorine-phosphate (FP) glasses $0.25Na_2O$-$0.5P_2O_5$-$0.05ZnF_2$-$0.1Ga_2O_3$-$0.05AlF_3$-$0.05NaF$ (mol%) doped with CdS, CdSe or PbSe (PbF_2+ZnS) and PbS (PbF_2+ZnSe) were synthesized The glass synthesis was conducted in an electric furnace at 1050°C in the Ar-atmosphere using the closed glassy carbon crucibles. This method allows for maintaining the high concentrations of volatile components (such as Se, Se ore F) and avoiding the interaction with atmospheric air, thus preventing the fluorine-oxygen substitution reaction. After quenching, the glasses were annealed at 320°C for 1 h to release thermal stress and to form MCs. Then glasses were cut into pieces of 10 × 10 mm and were optically polished. Planar polished samples 0.4–1.0 mm thick were prepared for further investigation. Samples were heat treated using a muffle furnace (Nabertherm) with program temperature control to induce formation of CdS(Se) and PbS(Se) QDs at a temperature above glass transition temperature (T_g).

3. Luminescent properties of quantum dots in the fluorine-phosphate glasses

3.1. Fluorine-phosphate glasses doped with CdSe quantum dots

The emission properties of semiconductor nanocrystals can be evaluated by three key parameters, which are the brightness, the emission color and the stability of the emission.

Samples of the glass-containing CdSe QDs were prepared by heat treatment of $0.25Na_2O$-$0.5P_2O_5$-$0.05ZnF_2$-$0.1Ga_2O_3$-$0.02PbF_2$-$0.08AlF_3$ glass doped with 0.6 mol% CdSe. The glass transition temperature measured with STA 449 F1 Jupiter (Netzsch) differential scanning calorimeter (DSC) was found to be 400 ± 3°C (**Figure 2**). DSC curve of the glass doped with CdS and CdSe was identical. High crystallization pick in 430–460 temperature range confirms high concentration CdSe(CdS) QDs in the glass. Hence, on the base of DSC data, we have determined the regimes of heat treatment of the starting glasses which are necessary for the nanocrystalline phase production: temperature $T = 410°C$, time 20–60 min.

The optical density spectra of the studied FP glass samples were recorded using a double-beam spectrophotometer Lambda 650 (Perkin Elmer) in the 1.5–5 eV spectral region with 0.1 nm resolution. Absorption spectra demonstrate that due to quantum size effect, the band gap of CdSe QDs increases from 2.2 to 3.0 eV, as the size of the nanocrystals decreases from 4.0 to 2.0 nm.

The emission color of the PL of the nanocrystals shifts continuously from red (centered at 730 nm) to orange (centered at 630 nm) as size of QDs decrease (**Figure 3**). QDs sizes were defined using data [34, 35].

In the PL spectra, the broad band with a large Stokes shift is dominant, and the band-edge PL is negligibly weak (**Figure 3**). The emission spectrum of samples is dominated by "deep trap" emission, strongly red shifted from the band edge (**Figures 3** and **4**). For registration of the emission spectra was used an EPP2000-UVN-SR (Stellar Net) fiber spectrometer. The luminescence was excited by semiconductor lasers (λ= 405 nm). **Figure 5** illustrates a luminescence shift from yellow to red with the increasing size of QDs.

Figure 6 demonstrates the excitation energy dependence of PL AQY magnitudes for glasses doped with CdSe QDs. The QDs concentration in the glasses 2 and 3 is equal (**Figure 3**), but AQY of the QD with size 3 nm is two times higher. The emission color of the PL of the QDs with size 3 nm is red with λ_{max} = 700 nm. Absolute quantum yield measurements were carried out inside the integrated sphere with Photonic Multichannel Analyzer (PMA-12, Hamamatsu) at room temperature. The measurement error for the absolute quantum yield was ±1%.

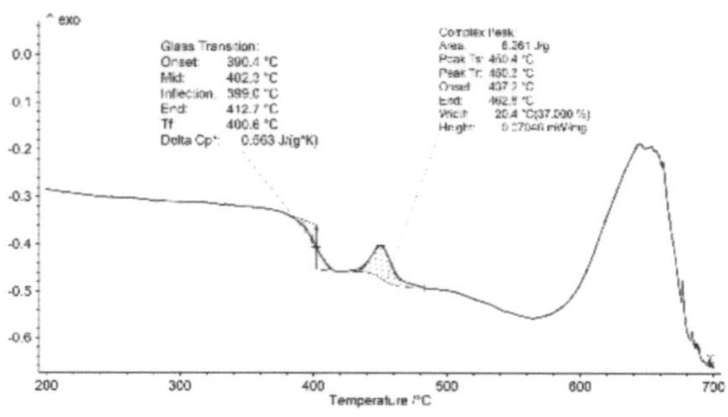

Figure 2. Thermal curve of the starting FP glass doped with Cd, Se and S measured with differential scanning calorimeter.

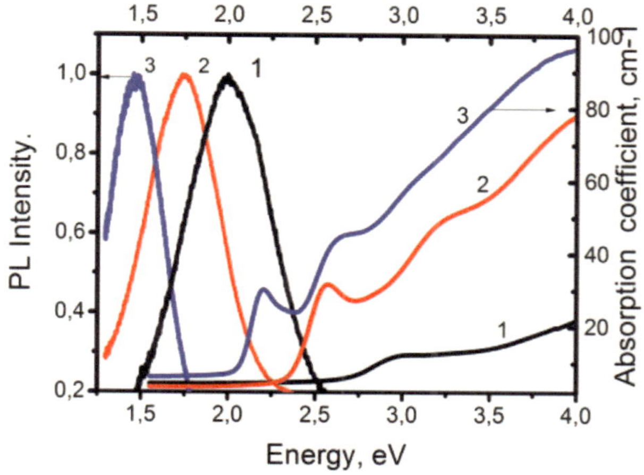

Figure 3. Absorption and luminescence spectra of the FP glass doped with CdSe QDs with sizes of 2.0 nm (1), 3.0 nm (2) and 4.0 nm (3). The excitation energy is 3.06 eV.

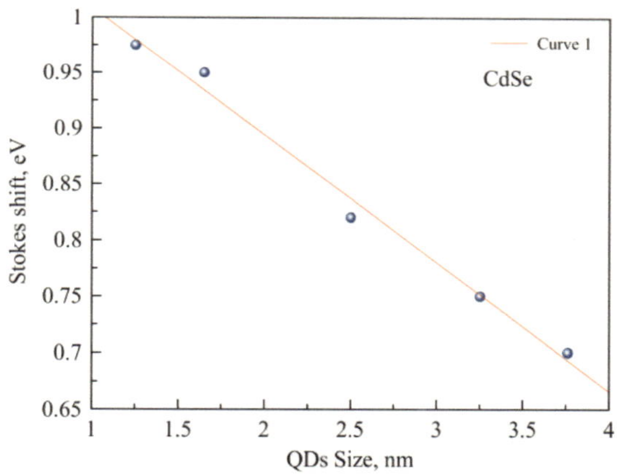

Figure 4. Dependence of the Stokes shift on the QDs size. Inset: Luminescence FP glasses doped with CdSe QDs with different sizes.

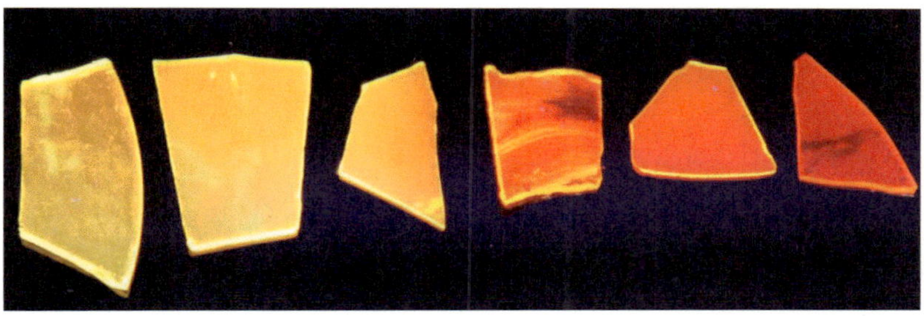

Figure 5. Photography of the luminescence of the glasses doped with CdSe QDs with sizes 2–4 nm.

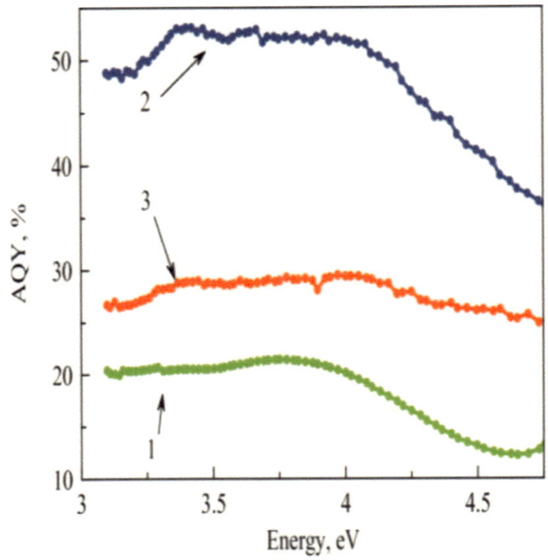

Figure 6. Dependence of the PL AQY for FP glasses doped with CdSe QDs with sizes 2.0 nm (1), 3.0 nm (2) and 4.0 nm (3) on the excitation energy.

The PL AQY magnitudes for glasses doped with CdSe QDs demonstrate a nonlinear dependence on the size (**Figure 7**). The PL AQY of the QDs increases monotonically from 4% to a maximum and then falls down to 30% (**Figure 4**). For convenience, the position with the maximum PL AQY is called the bright point as in Ref. [15]. The bright point for CdSe QDs in FP glass was observed for QDs with size 3.0 nm.

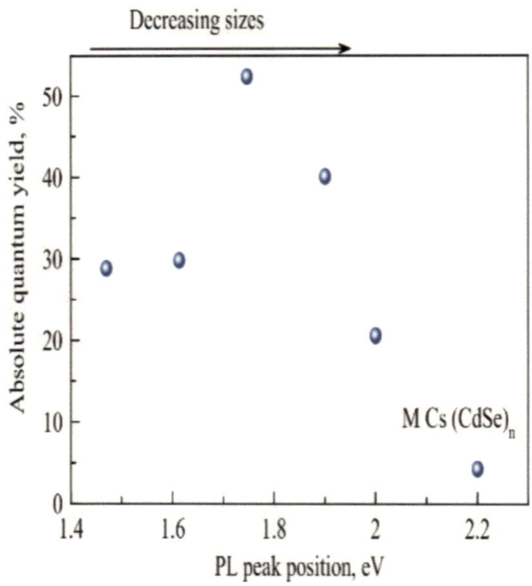

Figure 7. Dependence of the PL AQY on the PL peak position of the CdSe QDs.

The typical full width at half-maximum (FWHM) of the PL peak of the CdSe QDs ensemble at room temperature in FP glass, around 300–600 meV, is noticeably broader than that observed for colloidal QDs (30–70 meV) [15]. FWHM magnitudes decrease as QDs sizes increase (**Figure 8**).

3.2. Fluorine-phosphate glasses doped with CdS quantum dots

As a result of quantum confinement effects, the absorption bands of the colloidal semiconductors CdS QDs shift to the higher energy region compared with the band-gap energy of 2.5 eV in a CdS bulk crystal [15]. Heat treatment leads to this effect in the glasses. It was shown [36] that heat treatment has a significant impact on properties of glasses doped with CdS QDs. With increasing of the heat-treatment duration, the absorption band shifts to a lower energy side. These results indicate the formation and growth of the CdS QDs. Based on the theory of the quantum size effect in spherical QDs [37], the mean radii of prepared CdS QDs are estimated to be 2.3 and 3.5 nm. The observation of the clear absorption peaks indicates that the size-distribution width of the CdS QDs is rather small (**Figure 9**).

Figure 9 clearly shows effect of the heat treatment on the absorption and emission spectra of FP glasses doped with CdS QDs with sizes 2.3 and 3.5 nm. In PL spectra, the broad PL band with a large Stokes shift (1.2 eV) is dominant, and the band-edge PL is negligibly weak. The emission spectrum of samples is dominated by "deep trap" emission, strongly red shifted from the band edge.

Figure 9 demonstrates size dependence of the PL AQY magnitudes for glasses doped with CdS QDs with sizes 2.3, 3.5 and 1.5 nm.

The PL AQY magnitudes for glasses doped with CdS QDs with sizes 2.3–3.5 nm demonstrate weak dependence on the size (**Figure 10**). The PL AQY of the CdS QDs increases to 65% for QDs with size 2.3 nm and then slowly fells down to 60% for QDs with size 3.5 nm. In a whole range of the QDs sizes (1.8–3.5 nm) PL AQY magnitudes does not fall below 30%.

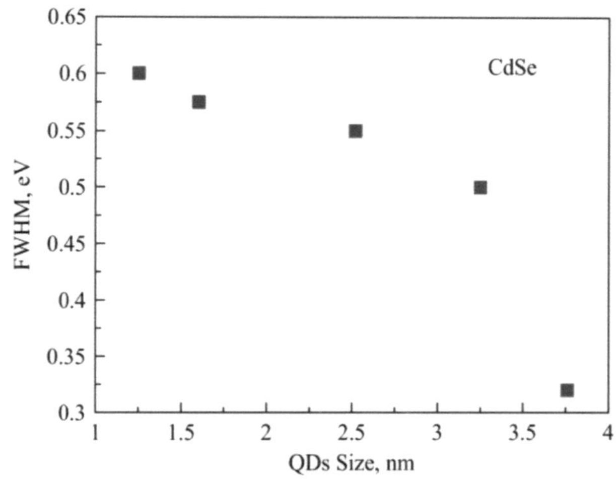

Figure 8. Dependence of the PL FWHM magnitudes for glasses doped with CdSe QDs with different sizes.

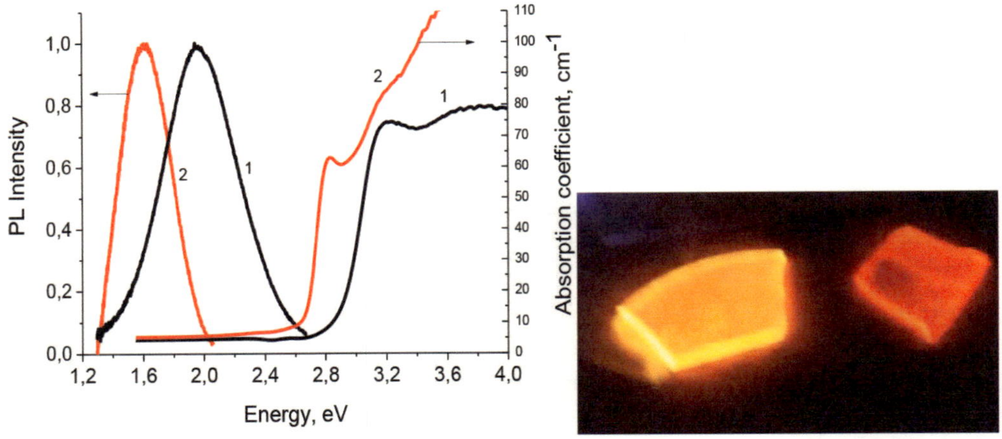

Figure 9. Absorption and luminescence spectra of the FP glasses doped with CdS QDs with sizes of 2.3 nm (1) and 3.5 nm (2). The excitation energy is 3.06 eV. Inset: Luminescence of the FP glasses doped with CdS QDs.

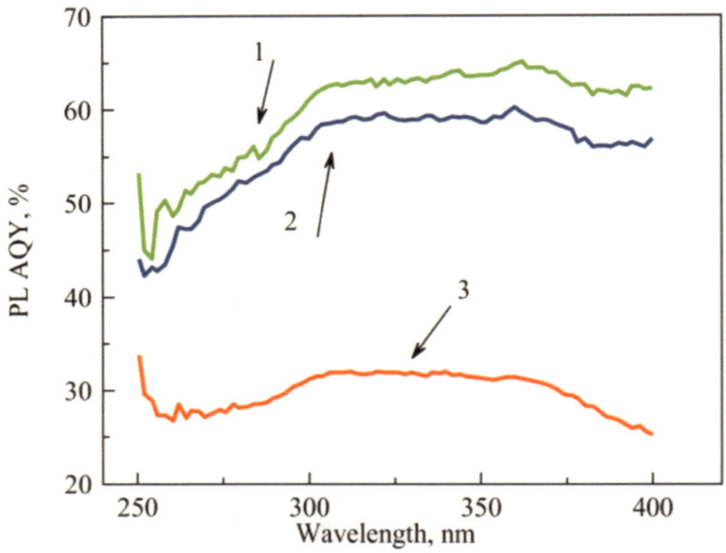

Figure 10. Dependence of PL AQY magnitudes for glasses doped with CdS QDs with sizes 2.3 nm (1), 3.5 nm (2) and 1.8 nm (3), respectively, on the excitation wavelength.

3.3. Stability of CdS and CdSe quantum dots in the fluorine-phosphate glasses

The PL properties of the CdS and CdSe nanocrystals prepared in the FP glasses, including the PL QY, the peak position and the PL FWHM, did not show any detectable change upon aging in air for several years.

It is known that under the various external factors (fluorescent lighting, laser and UV exposure in an oxygen environment) the CdS(Se) QDs sizes decrease [15] due to the formation of the layer in which sulfur or selenium substitute for oxygen. This process is named photo-oxidation. Photo-oxidation destroys the luminescent centers and reduces PL AQY magnitudes of the colloidal QDs.

A stability of nanocrystals in the FPG were defined by an impulse wave laser photo-oxidation experiment at UV (355 nm), visible (532 nm) wavelengths and after interrelation with Hg lamp during 3 h.

Figure 11 shows the result of the experiment on the FP glass samples doped with CdSe QDs in a diameter of 4.5 nm. The laser power density was 15 mJ and durations of the interaction laser change was from 0 to 15 min.

Before and after irradiation, the absorption spectra of the four samples were identical. This result demonstrates that the photo-oxidation of the nanocrystals in the glass is negligible.

A related photo-oxidation experiment was done on CdSe and CdS nanocrystals. The optical densities of the three samples were within 10% of each other below 375 nm. The CdS sample showed an absorption peak at 418 nm and for the CdSe sample, it is at 420 and 450 nm. A 240 W Hg lamp illuminated the three samples simultaneously for 3 h at a large enough distance from the lamp to prevent significant heating. As above, the CdS and CdSe nanocrystals did not show the blue shifting of their absorption maximums and a general loss of optical density at all wavelengths of absorption. The magnitudes of quantum yields were within the error of measurement for all tested glasses before and after the UV radiation.

Our qualitative results demonstrate the enhanced photostability of the QDs synthesized in the FP glasses. We measured the absorption spectra and the quantum yields of FP glasses doped with QDs before and after irradiation during 12 months. No changes in absorption spectra and in quantum yield were detected. This combination of photostability and high quantum yield makes these materials very attractive for use in optoelectronic devices like LEDs.

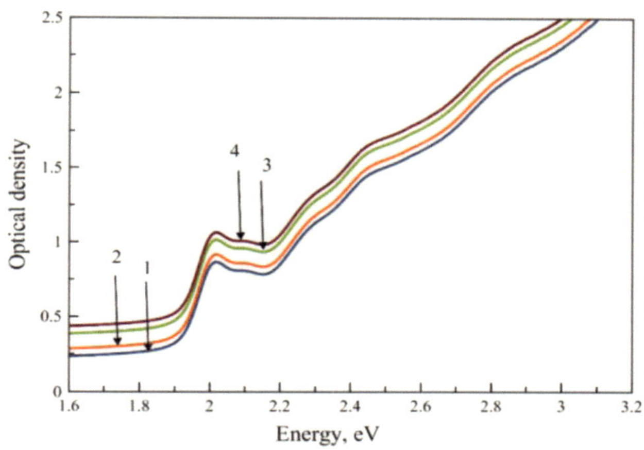

Figure 11. Absorption spectra of FP glasses doped with CdSe QDs with size of 4.5 nm after laser irradiation at power density 15 mJ during 0 min (1), 5 min (2), 10 min (3) and 15 min (4) at λ_{exc} = 355 nm. The diameter of the laser beam is 4 mm.

Absolute quantum yield allows to estimate the efficiency of converting UV light in the visible range and emphasizes the importance of this parameter for industrial applications of glasses doped with CdS(Se) QDs as luminescence down-shifting material or phosphor.

3.4. A fluorine-phosphate glass doped with CdS(Se) as a new type of red phosphor

The blue LED InGaN used as a basic component of the white LED emits with the spectra maximum at 450 nm. YAG:Ce3+ powder is a conventional phosphor with the spectral maximum at 570 nm and shifts the LED color temperature range toward the "warmer" white light. However, this composition gives the so-called "cold" white light, because their radiation does not cover the overall visible range. This "cold" white light is not always comfortable for eyes. Thus if one wants to get the composition of the emitters with "warm" white light, it is necessary to introduce an additional phosphor—some new component, introducing into the spectrum the red component with broad band luminescence and the maximum at 600–750 nm. Prospective for this task are, for instance, materials doped with manganese ions. However, this type of LEDs has a significant disadvantage—the absorption cross-section by manganese ions in the spectral range of 400–600 nm is negligible.

One can solve this problem with the use of other phosphor activators—CdS(Se) QDs with the maximum of the emission band at 650–700 nm, which introduces into the blue LED emission of the red component. As a result, the mixing of blue, green and red light takes place. In our work we have tried to solve the mentioned problems by means of the FP glasses doped with CdS(Se) QDs.

Figure 12 shows the CIE chromaticity coordinates of the FP(CdSe) glass heat treated at $T = 405°C$ for 30 and 60 min and **Figure 13** illustrates CIE chromaticity coordinates of the FP(CdS) glass heat treated at $T = 415°C$ for10, 30 and 40 min. It is shown the increasing of the heat-treatment duration shifts the chromaticity coordinates from the yellow area (x = 410, y = 370) to the area of the diagram that corresponds to the red color (x = 440, y = 380).

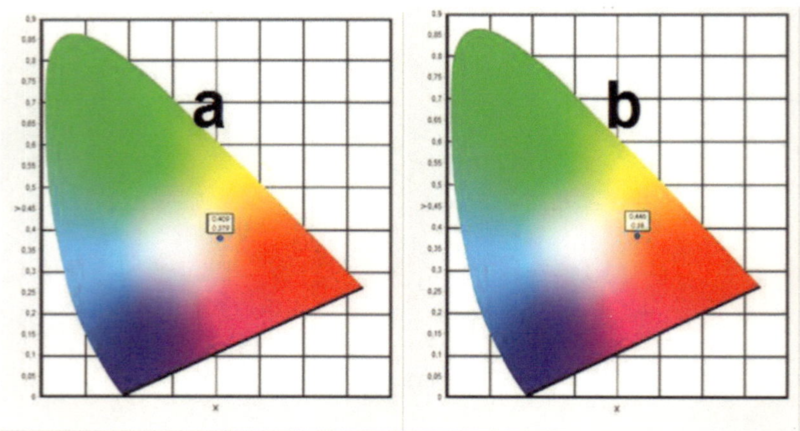

Figure 12. CIE chromaticity coordinates of the FP(CdSe) glass heat treated at $T = 405°C$ during 30 min (a) and 60 min (b).

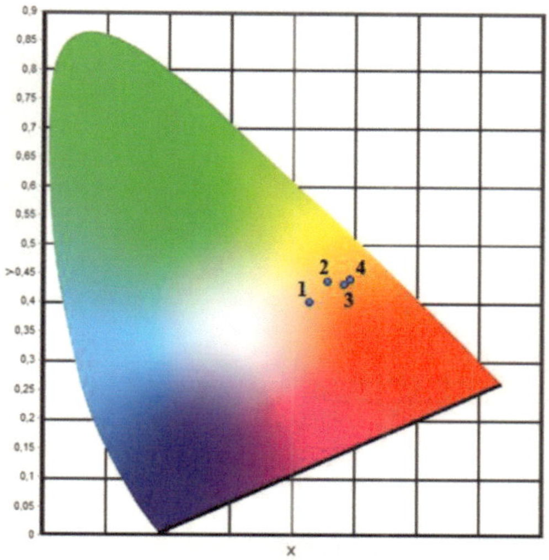

Figure 13. CIE chromaticity coordinates of the FP (CdS) glass heat treated at $T = 415°C$ during 10 min (1), 30 min (2), 40 min (3) and 60 min (4).

In this study, a new type of red phosphor representing a fluorine-phosphate glass doped with CdS(Se) were synthesized using technology developed and applied earlier by Vaynberg [1]. These red phosphors with combination of green-yellow phosphors can be applied to fabricate the pc-WLEDs on the blue InGaN chips.

Absolute quantum yield allows estimating efficiency of converting UV light in the visible range that is why it is an important parameter for industrial applications of glasses doped with CdS(Se) QDs as luminescence down-shifting material or phosphor.

3.5. Fluorine-phosphate glasses doped with PbS and PbSe molecular clusters

Highly concentrated semiconductor components, which can be stored in glasses, is a main difference between FP glasses and previously studied silicate and borosilicate glasses. It is obviously expressed, that the concentration of MCs and QDs, which may be formed in the glass during heat treatment, is also significantly higher. High concentration and the uniform distribution of the activators should lead to high values of absolute quantum yield. DSC curves confirm the high concentration of the activator (**Figure 14**).

The glass transition temperature was found to be 400–408 ± 3°C. Samples were heat treated to induce formation of PbS(Se) QDs at 410°C. Due to the high concentration of sulfur and selenium, crystallization peak induced by the precipitation of PbS(Se) QDs was observed.

As-prepared glasses doped with PbS(Se) demonstrate strong luminescence (when excited in UV spectral region) due to the MCs formation whereas the changes of the absorption coefficient are not large. The heat treatment at temperatures less than T_g results in the MCs' growth and increases the luminescence intensity [36–38]. For registration of the emission spectra excited at $\lambda = 405$ nm (3.06 eV), we used an EPP2000-UVN-SR (Stellar Net) fiber spectrometer.

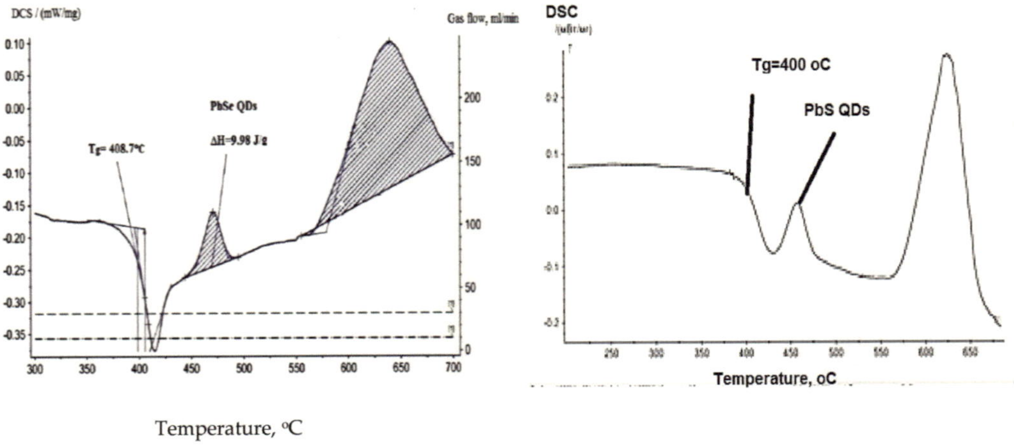

Figure 14. DSC curves of the initial glasses doped with PbSe (a) and PbS (b).

The optical density spectra of the studied FP glass samples were recorded using a double-beam spectrophotometer Lambda 650. (Perkin Elmer) in the 1.5–5 eV spectral region with 0.1 nm resolution and spectrophotometer Cary 500 (Varian) from 300 to 3000 nm (optical resolution was 0.5 nm).

PL excitation spectrum of the glass doped with $MCs(PbS)_n$ can be observed in spectral range of 250–400 nm (**Figure 15a**). Asymmetric form of the PL excitation band by two (λ_{max} = 290 nm and λ_{max} = 350 nm) indicates the existence of various sizes of the MCs. **Figure 15b** shows luminescence spectra of $MCs(PbS)_n$. The broad luminescence band of the MCs shifts from 570 to 700 nm due to the MCs sizes increasing. The absolute quantum yield of the as-prepared glasses doped with PbS (1) and after heat treatment (2) are shown in **Figure 16a**. The absolute

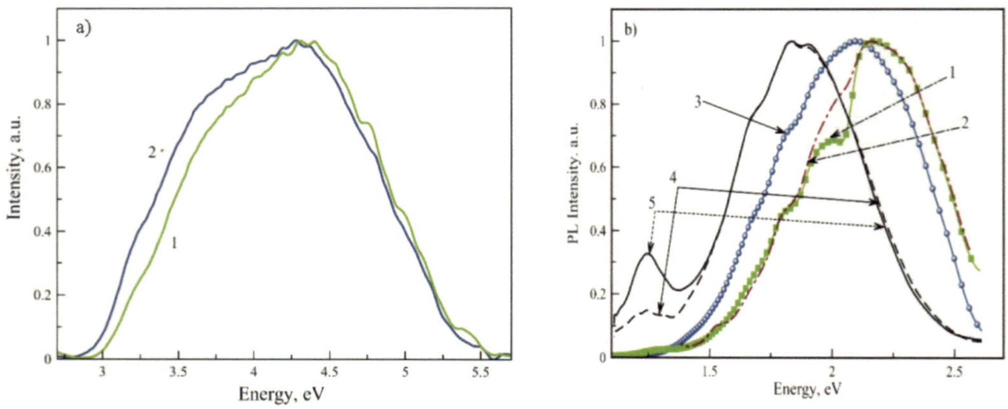

Figure 15. PL excitation spectrum of the glass doped with MCs $(PbS)_n$ (λ_{lum} = 540 nm) (a) Luminescence spectra of the glasses doped with MCs $(PbS)_n$: as-prepared (1) and after heat treatment at T_{HT} = 410°C within 10 min (2), 20 min (3), 30 min (4) and 40 min (5) (λ_{exc} = 250 nm) (b).

quantum yield of the as-prepared glass doped with PbSe (1) and after heat treatment (2) shown in **Figure 16b**. **Figure 16c** shows the photography of the $(PbS)_n$ MCs luminescence at $\lambda_{exc} = 365$ nm. Absolute quantum yield of the FP glass doped with PbS and PbSe MCs depends on excitation wavelength and changes from 10 to 2% (**Figure 16a** and **b**). The MCs are nucleation centers for the formation of PbS and PbSe QDs. It was shown [33, 36] that QDs formation in FP glasses leads to an increase in its intensity by a factor of five (**Figure 16a** and **b**).

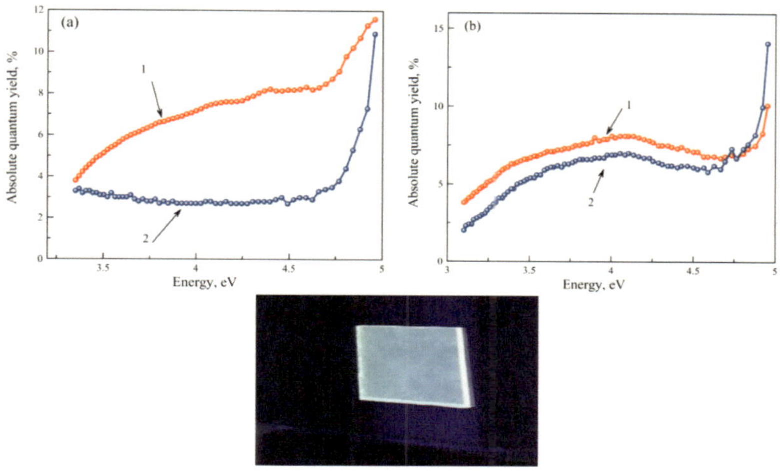

Figure 16. (a) Absolute quantum yield for the glasses doped with PbS as-prepared (1) and glass after heat treatment (2), (b) Absolute quantum yield for glass doped with PbSe as-prepared (1) and after heat treatment (2) and (c) photography of the $(PbS)_n$ MCs luminescence at $\lambda_{exc} = 365$ nm.

The luminescence spectrum of FP (PbS) after heat treatment at $T = 360°C$ within 90 min (1) (**Figure 17a**) demonstrates appearance of the second band (980 nm) due to small QDs formation. After heat treatment within 97 min (2) the luminescence band of the MCs was disappeared (**Figure 17a**).

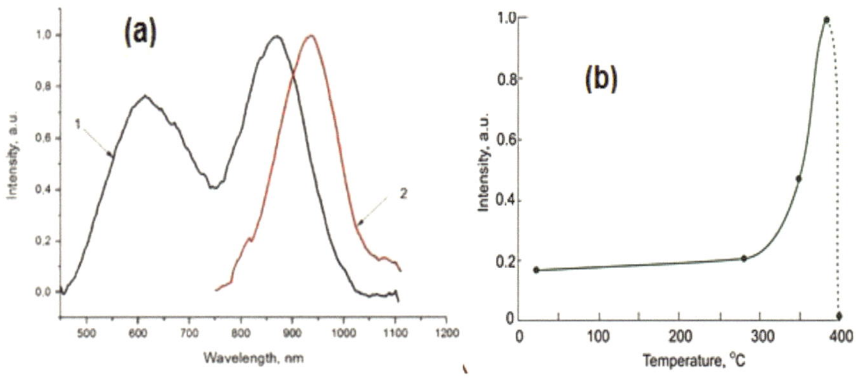

Figure 17. (a) Luminescence spectra of FP glass doped with MCs $(PbS)_n$ after heat treatment at 400°C within 90 min (1) and 97 min (2), $E_{ex} = 3.06$ eV and (b) Luminescence intensity in maximum as a function of thermal treatment temperature.

Figure 18 shows the influence of heating on luminescence spectra. In this experiment, the sample preliminary thermal treated at 380°C was used. The increase of the sample temperature from 20 to 250°C led to seven times decrease of luminescence intensity and to the weak red shift of the luminescence band. This effect is reversible and can be repeated for many times [36]. The luminescence thermal quenching can be explained by a fact that with the temperature rise the electrons in the excited state can occupy high vibrational energy levels, which intersect the ground state level at configuration coordinate diagram. This allows the vibrational relaxation of the excited electrons to the ground state via phonon release without emission of radiation [36].

3.6. Fluorine-phosphate glasses doped with PbS and PbSe quantum dots

For luminescence measurement (scanning ranges 400–1100 nm and 900–1800 nm) with fixed excitation wavelength (λ_{ex} = 808 nm) Stellar Net EPP2000-UVN-SR fiber spectrometer was used (optical resolution 0.5 nm). A semiconductor laser with pumping power 0.1 W excited the luminescence.

When heat-treatment duration or temperature increase, the new band appears with λ_{max} = 1000 nm. It is due to the PbSe (S) QD formation with the sizes below 2.0 nm (**Figure 16a**). Increase of the heating temperature and time results in a prominent change of the absorption spectra and appearance of discreet spectra corresponding to QDs.

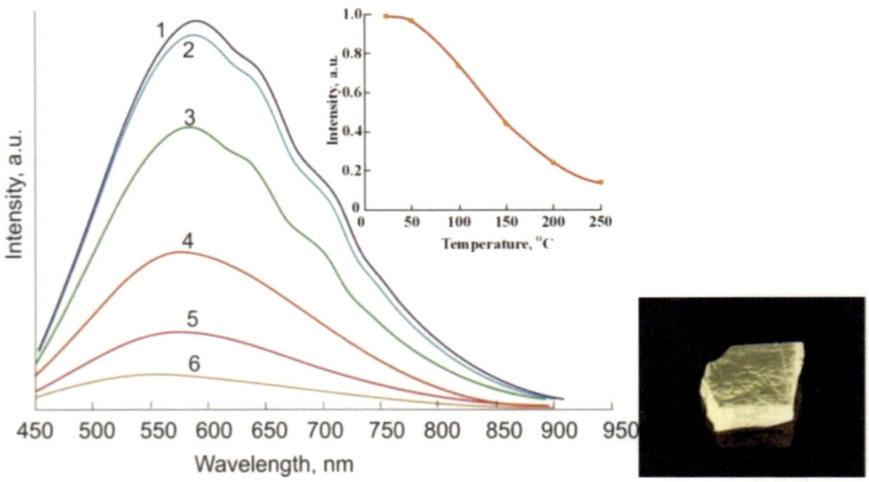

Figure 18. The influence of the temperature of the FP glasses doped with $(PbSe)_n$ MCs on the luminescence intensity..- The luminescence spectra were measured at temperature of (1) 22°C, (2) 50°C, (3) 100°C, (4) 150°C, (5) 200°C and (6) 250°C. Excitation wavelength was of 405 nm. Inset: Luminescence intensity in maximum via the temperature of the sample and photo of the luminescence of the FP glass doped with $(PbSe)_n$ MCs.

To precipitate PbS QDs, glass samples were heat treated at temperature 410°C within 20–90 min (**Figure 19**). Based on hyperbolic band model obtained from $k \times p$ calculations [21], average diameters of these PbS QDs were found to be 3.0, 3.9 and 4.9 nm. QDs sizes were calculated using formula:

$$E_0 = 0.41 + \frac{1}{0.0252 \cdot D^2 + 0.283 \cdot D} \quad (1)$$

where E_0—energy gap of PbS QDs (eV) and D—diameter of QDs in nm.

Increasing heat-treatment time and temperature leads to shifting of luminescence bands of the PbS QDs to 970, 1300 and 1500 nm (**Figure 19**) and absorption bands (to 920, 1100 and 1400 nm). Stokes shift changes from 80 to 50 meV due to the changing of QDs sizes from 3.0 to 4.9 nm (**Figure 21b**).

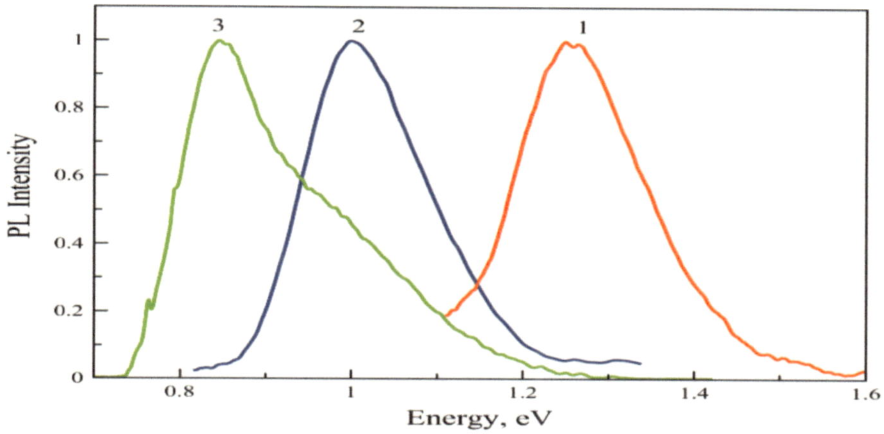

Figure 19. Luminescence spectra of the glasses doped with PbS QDs with sizes 3 nm (1), 3.9 nm (2), 4.9 nm (3) and λ_{exc} = 808 nm.

Figure 20 shows absorption and luminescence spectra of PbSe QDs formed in glasses. Increasing of the heating time results in prominent changes of absorption spectra of the FP glass doped with PbSe and leads to appearance of discreet spectra corresponding to QDs. Based on hyperbolic band model obtained from tight binding calculation using experimental energy values E_g, we can estimate the PbSe QD sizes [39] according to Eq. (2),

$$E_g(D) = E_g(\infty) + \frac{1}{0.0105 \cdot D^2 + 0.2655 \cdot D + 0.0667} \quad (2)$$

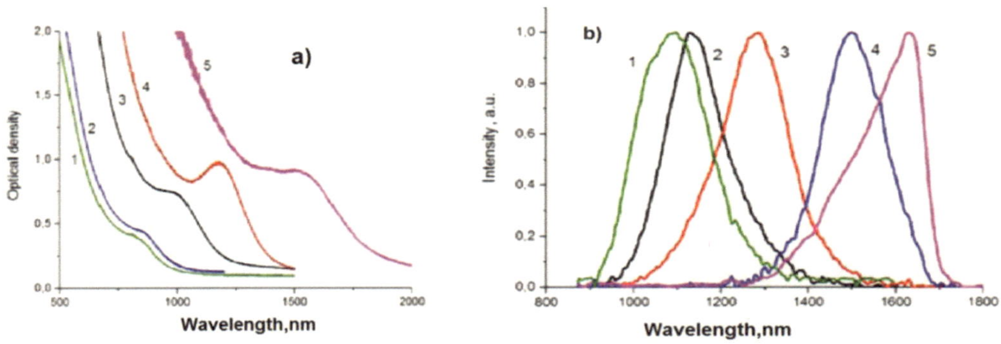

Figure 20. Absorption (a) and luminescence (b) spectra of the glasses doped with PbSe QDs with sizes 2.5 nm (1), 3.0 nm (2), 3.7 nm (4), 5.1 nm (5), λ_{exc} = 808 nm.

where D is the effective diameter of the quantum dots (nm), E_g is the energy gap of PbSe quantum dots (eV) and $E_g(\infty)$ is the energy bandgap of bulk PbSe semiconductor (0.29 eV).

When heat-treatment time and temperature are increased, absorption peaks shifts to 870–1540 nm and luminescence bands shifts to 1050–1650 nm. Stokes shift of QDs changes from 335 to 60 meV when size varies from 2.5 to 5.1 nm (**Figure 21a**).

It can be deduced that heat-treatment conditions define the formation characteristic of PbS (Se) QDs in glass (such as the beginning growth temperature, growth rate, density in glass matrix, size and size distribution of QDs, etc.).

Because of matching of the PL characteristic with window (1330 nm) of PbSe QD-embedded glass (see the **Figure 4**, curve 3), we choose this glass as the representative sample to characterize the optical amplification at the 1330 nm window. Signals at 1330 nm with different

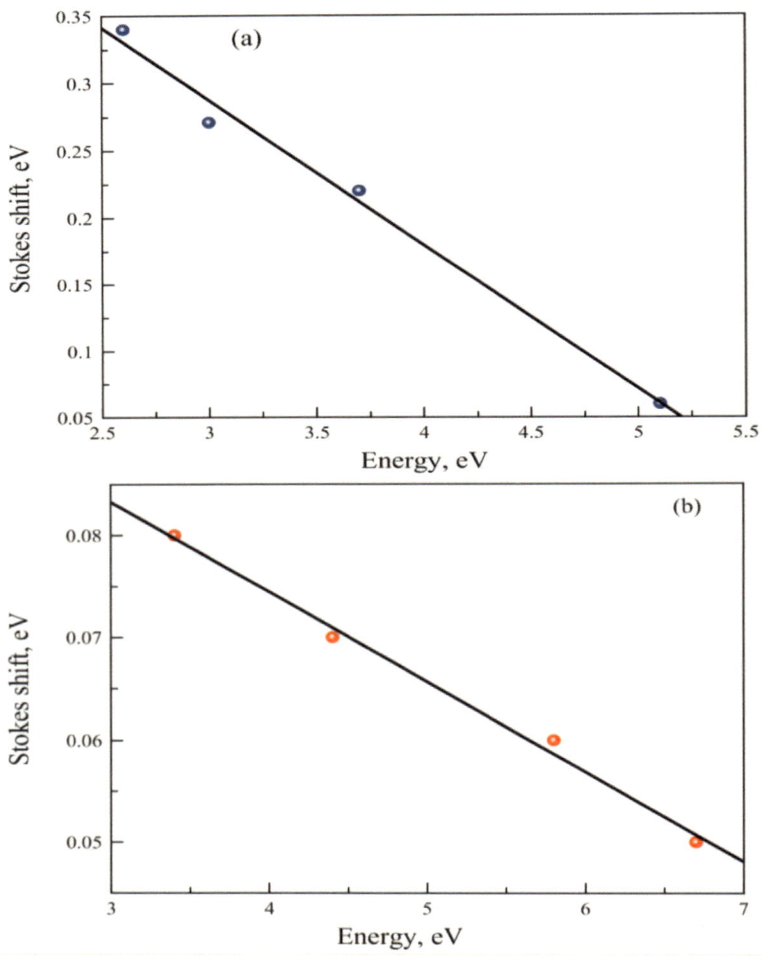

Figure 21. Stokes shift versus QDs diameter for PbS (a) and PbSe (b) QDs.

http://dx.doi.org/10.5772/intechopen.68459

pump power values were measured. It can be found that the intensity enhances gradually with increasing of power. Even when the pump power reaches from 500 to 1000 mW, no obvious signal saturation is detected. This allows that FP (PbS and PbS) glasses can be potentially applied in broadband amplifiers and confirm the high photo-stability of the QDs synthesized in the FP glasses. We also measured the absorption spectra and quantum yields of FP glasses doped with QDs before and after irradiation during 12 months. No changes in absorption spectra and in quantum yield were detected.

The FP glasses doped with PbSe and PbS QDs are infrared fluorophores, which are efficient, photo-stable and possess short lifetimes. These materials may find unique applications for fluorescent imaging.

4. Conclusions

The CdS and CdSe nanocrystals synthesized in the fluorine-phosphate glass represents a series of excellent emitters in the orange-red spectral region (600–750 nm) in terms of their PL AQY and the FWHM of the PL spectra, and they show the photo- and chemical stability of the emission for a long time.

The photoluminescence quantum yield of CdSe QDs rises monotonically to a maximum value and then decreases gradually with the size increase of QDs. Such a maximum (a PL "bright point") is in 650–750 nm spectral range.

The PL AQY magnitudes for glasses doped with CdS QDs with sizes of 2.3–3.5 nm demonstrate weak dependence on the size and reaches 65%.

We suggest that origin of these dependences is the difference in the interaction mechanisms between CdSe and CdS quantum dots and glass network.

Experimental results suggest that the existence of the PL bright point is a general phenomenon of CdSe QDs and similarly is a signature of an optimal surface structure reconstruction of the nanocrystals grown in a liquid or in a glass. Absolute quantum yield magnitude of luminescence glasses doped with CdS (Se) QDs can reach 50–65%, which is two times higher than it was reported earlier in silicate glasses. It opens up new prospects for using such materials as phosphors for white LEDs and down-convertors for solar cells.

It was shown that heat treatment of the FP glasses leads to formation of $(PbS)_n$ and $(PbSe)_n$ molecular clusters, which exhibit luminescent properties in visible range with quantum efficiency from 2 to 10%. Increasing the heat-treatment temperature results in the PbS and PbSe QDs (sizes of 3–5 nm) formation with high concentration (~1 mol%). The QDs have broadband photoluminescence with quantum efficiency about five times more than MCs (~50%) in the spectral range of 1–1.7 μm. FP glasses doped with PbSe and PbS QDs are infrared fluorophores, which are both efficient and possess short lifetimes. These materials may find unique applications for fluorescent imaging tagging in the near-IR spectral range or as fluorescent materials in the fiber communication range of 1.3–1.5 microns.

Acknowledgements

Research was funded by Russian Science Foundation (Agreement #14-23-00136).

Author details

Elena Kolobkova[1,2]* and Nikolay Nikonorov[1]

*Address all correspondence to: kolobok106@rambler.ru

1 Department of Optical Informatics Technologies and Materials, ITMO University, Saint-Petersburg, Russia

2 St. Petersburg State Institute of Technology (Technical University), St. Petersburg, Russia

References

[1] Vaynberg B, et al. High optical nonlinearity of CdS_xSe_{1-x} microcrystals in fluorine-phosphate glass. Optics communications. 1996;**132**(3-4):307-310

[2] Guerreiro PT, Ten S, Borrelli NF, Butty J, Jabbour GE, Peyghambarian N. PbS quantum-dot doped glasses as saturable absorbers for mode locking of a Cr: forsterite laser. Applied Physics Letters. 1997;**71**:1595-1597

[3] Heo J, Liu C. PbS quantum-dots in glass matrix for universal fiber-optic amplifier. Journal of Materials Science: Materials in Electronics. 2007;**18:**S135-S139

[4] Borreli NF, et al. Quantum confinement effects of semiconducting microcrystallites in glass. Journal of Applied Physics. 1987;**61**:5399-5409

[5] Su Z, et al. Selenium molecules and their possible role in deep emission from glasses doped with selenide nanocrystals. Journal of Applied Physics. 1996;**80**:1054-1055

[6] Xu KM, et al. Optical properties of CdSe quantum dots in silicate glasses. Journal of Non-Crystalline Solids. 2010;**356**:2299-2301

[7] Loiko PA, Rachkovskaya GE, Zaharevich GD, Gurin VS, Gaponenko MC, Yumashev KV. Optical properties of novel PbS and PbSe quantum dot-doped alumino-alkali-silicate glasses, Journal of Non-Crystalline Solids. 2012;**358:**1840-1845

[8] Han M, et al. Quantum-dot-tagged microbeads for multiplexed optical coding of biomolecules. Nature Biotechnology. 2001;**19**:631-635

[9] Bruchez M, et al. Semiconductor nanocrystals as fluorescent biological labels. Science. 1998;**281**:2013-2016

[10] Chan WCW, Nie SM. Quantum dot bioconjugates for ultrasensitive nonisotopic detection. Science. 1998;**281**:2016-2018

[11] Artemyev M, et al. Light trapped in a photonic dot: Microspheres act as a cavity for quantum dot emission. Nano Letters. 2001;**1**:309-314

[12] Klimov VI, et al. Optical gain and stimulated emission in nanocrystal. Science. 2000;**290**: 314-317

[13] Sundar VC, et al. Full color emission from II-VI semiconductor quantum dot-polymer composites. Advance Materials. 2000;**12**:1311-1312; Novel light emitting devices using cadmium selenide nanocrystals. Abstracts of papers of the American chemical society. 220. pp. U206-U206.

[14] Schlamp MC, Peng X, Alivisatos AP. Improved efficiencies in light emitting diodes made with CdSe(CdS) core/shell type nanocrystals and a semiconducting polymer. Journal of Applied Physics. 1997;**82**:5837-5842

[15] Qu L, Peng X. Control of photoluminescence properties of CdSe nanocrystals in growth. Journal of American Chemical Society. 2002;**124**(9):249-255

[16] Fu H, Zunger A. InP quantum dots: Electronic structure, surface effects, and the red-shifted emission. Physical Review B: Condensed Matter and Materials Physics. 1997;**56**: 1496-1508

[17] Talapin DV, et al. Highly luminescent monodisperse CdSe and CdSe/ZnS nanocrystals synthesized in a hexadecylamine-trioctylphosphine oxide-trioctylphospine mixture. Nano Letters. 2001;**14**:207-211

[18] Wang X, et al. Surface-related Eemission in highly luminescent CdSe quantum dots. Nano Letters. 2003;**3**(8):1103-1106

[19] Kim JM, et al. Photoluminescence enhancement in CdS quantum dots by thermal annealing. Nanoscale Research Letters. 2012;**7**:482-489

[20] Kang FW. Wise, Electronic structure and optical properties of PbS and PbSe quantum dots. The Journal of the Optical Society of America B. 1997;**14**:1632-1646

[21] Wisem FW. Lead salt quantum dots: The limit of strong quantum confinement. Accounts of Chemical Research. 2000;**33**:773-780

[22] Sargent EH. Infrared quantum dots. Advanced Materials. 2005;**17**:515-522

[23] Akiyama T, Kuwatsuka H, Simoyama T, Nakata Y, Mukai K, Sugawara M, Wada O, Ishikawa H. Nonlinear gain dynamics in quantum-dot optical amplifiers and its application to optical communication devices. The IEEE Journal of Quantum Electronics. 2001;**37**:1059-1065

[24] Schaller RD, Petruska MA, Klimov VI. Tunable near-infrared optical gain and amplified Sspontaneous emission using PbSe nanocrystals. The Journal of Physical Chemistry B. 2003;**107**:13765-13768

[25] Pang F, Sun X, Guo H, Yan J, Wang J, Zeng X, Chen Z, Wang T. A PbS quantum dots fiber amplifier excited by evanescent wave. Optics Express. 2010;**18**:14024-14030

[26] Dong G, Wu G, Fan S, Zhang F, Zhang Y, Wu B, Ma Z, Peng M, Qiu J. Formation, near-infrared luminescence and multi-wavelength optical amplification of PbS quantum dot-embedded silicate glasses. Journal of Non-Crystalline Solids. 2014;**383**:192-195

[27] Moreals I, Lambert K, Smeets D, Muynck DD, Nollet T, Martins JC, Vanhaecke F, Vantomme A, Delerue C, Allan G, Hens Z. Size-dependent optical properties of colloidal PbS quantum dots. ACS Nano. 2009;**3**:3023-3030

[28] Kolobkova EV, Lipovskii AA, Petrikov VD. Fluorophosphate glasses containing PbSe quantum dots. Glass Physics and Chemistry. 2002;**28**:246-250

[29] Lipovskii AA, et al. Formation and growth of semiconductor nanocrystals in phosphate glass matrix. Journal of the European ceramic society. 1999;**19**(6-7):865-869

[30] Dantas NO, Silva RS, Qu F. Optical properties of PbSe and PbS quantum dots embedded in oxide glass. Physica Status Solidi B. 2002;**232**:177-181

[31] Kolobkova EV, Lipovskii AA, Petrikov VD, Melekhin VG. Fluorophosphate glasses with quantum dots based on lead sulfide. Glass Physics and Chemistry. 2002;**28**:251-255

[32] Martin JL, Rivera R, Cruz SA. Confinement of excitons in spherical quantum dots. Journal of Physics: Condensed Matter. 1998;**10**:1349-1361

[33] Lipatova ZO, Kolobkova EV, AseevVA. Kinetics and luminescence of cadmium sulfide quantum dots in fluorine-phosphate glasses. Optics and Spectroscopy. 2015;**119**(I2): 229-233

[34] Norris D, Bawendi MG. Measurement and assignment of the size-dependent optical spectrum in CdSe quantum dots. Physical Review B. 1996;**53**(24):16338-16446

[35] Norris DJ, Bawendi MG. Structure in the lowest absorption feature of CdSe quantum dots. The Journal of Chemical Physics. 1995;**103**(130):5260-5268

[36] Kolobkova EV, Kukushkin DS, Nikonorov NV, Shakhverdov TA, Sidorov AI, Vasiliev VN. Luminescent properties of fluorophosphate glasses with lead chalcogenides molecular clusters. Journal of Luminescence. 2015;**162**:36-40

[37] Moller K, Bein T, Herron N et al. Encapsulation of lead sulfide molecular clusters into solid matrices: Structural-analysis with x-ray absorption-spectroscopy. Inorganic chemistry. 1989;**28**:2914-2919.

[38] Evans CM, Guo L, Peterson JJ, Maccagnano-Zacher S, Krauss TD. Ultrabright PbSe magic-sized clusters. Nano Letters. 2008;**8**:2896-2899

[39] Allan G, Delerue C. Confinement effects in PbSe quantum wells and nanocrystals, Physical Review B. 2004;**70**:245321

Section 2

Semiconductor Quantum Light Sources

Chapter 6

Quantum Dot–Incorporated Hybrid Light-Emitting Diodes

Namig Hasanov

Additional information is available at the end of the chapter

http://dx.doi.org/10.5772/intechopen.68356

Abstract

Quantum dots are very promising candidates to enhance the performance of hybrid devices. Their size-dependent wavelength tunability owing to quantum size effect, narrow full width at half maximum, high quantum yield, and several other optoelectronic properties enable their use as potential components in GaN-based light-emitting diodes. This chapter explains methods to fabricate color-converted and white light-emitting diodes with the incorporation of semiconductor quantum dots.

Keywords: quantum dots, light-emitting diodes, color conversion, white light

1. Introduction

Thanks to their physical, optical, and electronic properties, semiconductor nanocrystals attracted enormous interest as promising nanomaterials with potential applications in optoelectronics [1]. These nanocrystals can be chemically synthesized with different sizes and shapes including nanowires, nanodiscs, and quantum dots (QDs). Spherical QDs possess several advantages that make them more flexible nanomaterials to be utilized in wider range of applications. QDs can be synthesized as a core, core-shell, or core-multi shell as shown in **Figure 1**. The size of QDs determines their emission wavelength; emission energy of these nanocrystals is related to their quantum confinement property that changes with the radius of the nanocrystals. Thus, it is possible to synthesize nanocrystals emitting with wavelength range covering the whole visible spectrum. This property makes them powerful optoelectronic components. The full width at half maximum (FWHM) of semiconductor QDs is generally in the range of 30–40 nm. Recently, QDs with very high quantum efficiency (one of the main performance measures of QDs) values were reported.

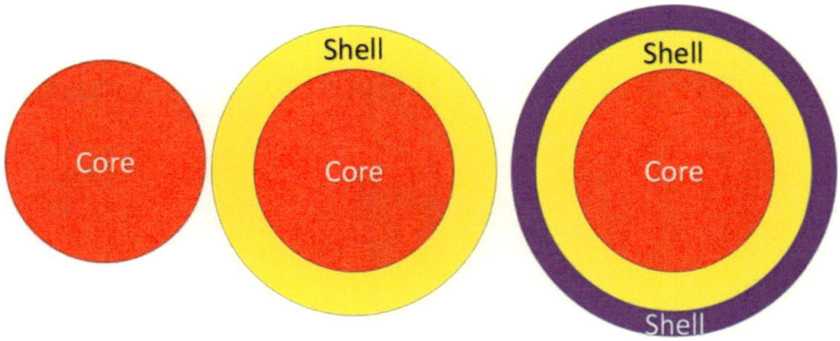

Figure 1. Schematic diagram showing core, core-shell, and core-multi shell QDs.

Although, QDs exhibit high optical properties, their electrical properties are not at the desired level. Thus, the efficiency of the devices utilizing electrically injected QDs is very low [2]. One of the main reasons for this low performance of QD-incorporated devices is the existence of organic ligands on the surface of QDs that prevents an efficient current injection. On the other hand, ligands play an important role in the enhancement of the stability of QDs. Further research is needed to be done to improve the electrical injection properties of these semiconductor nanocrystals. However, their optical properties can be used to enhance the performance of the devices with lower optical outputs. In the following sections, the device structures incorporating QDs are introduced and their effectiveness is thoroughly discussed.

2. Color-converter QDs on GaN-based devices

Since the important inventions mainly by Nakamura, Akaşaki, and Amano in the development of GaN-based light-emitting diodes (LEDs) (they received Nobel Prize in Physics for the invention of blue LED in 2014) [3–5], these devices were widely investigated to enhance the external quantum efficiency and optical power as well as to reduce the electrical injection issues [6, 7]. GaN-based LEDs are mostly grown with metalorganic chemical vapor deposition (MOCVD) on c-plane sapphire substrates. The main drawbacks of sapphire as a growth substrate are the lattice mismatch and thermal expansion coefficient mismatch. The former drawback significantly reduces the performance of devices due to the formation of strain in the epitaxial layers. However, the performance of LEDs grown on sapphire is still higher than those grown on other substrates [8]. The higher efficiency can be achieved in the LEDs incorporating InGaN quantum wells in between the p-type and n-type GaN epitaxial layers. **Figure 2** depicts a schematic structure of a typical multiple quantum well InGaN/GaN LED. To achieve blue emission, the special care needs to be taken during the growth of InGaN quantum wells as the emission is mainly realized through the recombination of electrons and holes in these wells. The amount of In incorporated during the growth defines the emission wavelength of the device. Thus, it is possible to achieve high-quality blue emission with the incorporation of the correct amount of In (usually around 15%) [9]. Photoluminescence from an epitaxially grown blue LED is shown in **Figure 3**. In order to fabricate a GaN-based

http://dx.doi.org/10.5772/intechopen.68356

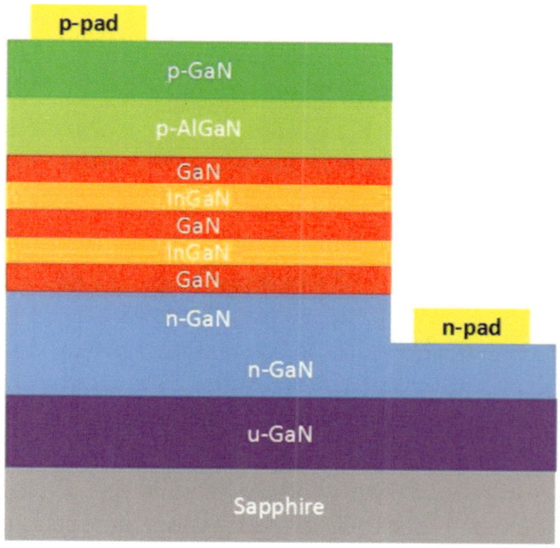

Figure 2. Schematic diagram of multiple quantum well InGaN/GaN LEDs.

device emitting light with longer wavelength, a larger amount of In should be introduced into the InGaN compound layer. However, the growth of InGaN layer with larger amounts of In results in the segregation of In [10]. Thus, InGaN/GaN LEDs utilizing more In exhibit significantly lower optical power and external quantum efficiency when compared with the blue LEDs using smaller amounts of In.

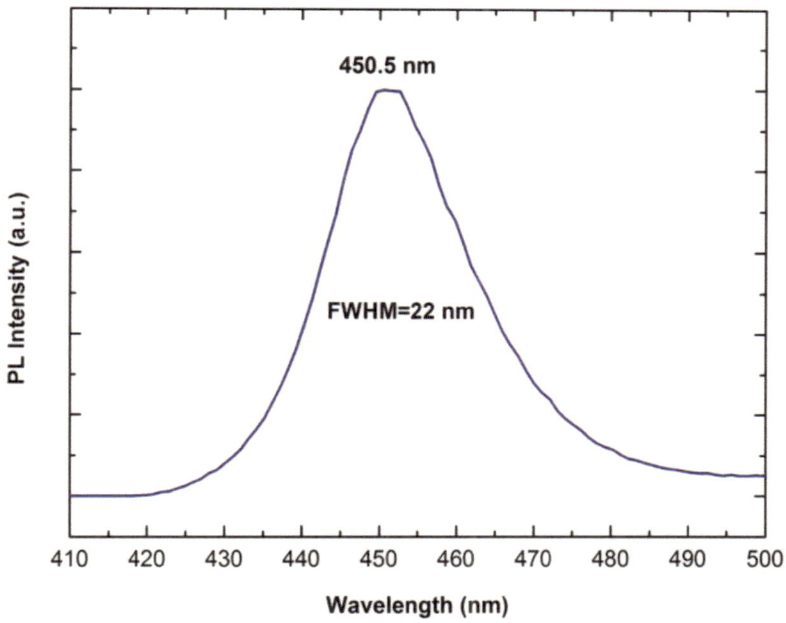

Figure 3. Photoluminescence of an epitaxially grown blue multiple quantum well InGaN/GaN LED.

To enhance the output performance of InGaN/GaN LEDs emitting the longer wavelength range, QDs can be incorporated as color-converter components [11]. Thus, by utilizing the optical properties of QDs and electrical properties of InGaN/GaN LED structure, it is possible to fabricate a color-converted hybrid LED emitting at longer wavelength. To realize this kind of device, QDs are placed on top surface of InGaN/GaN LED structures. One of the critical points during the construction is to place the correct amount of QDs. As the emission wavelength of the LED and the QDs are different, the process of placing QDs on top can result in either full color conversion or mixed color emission. Thus, the optimized amount of QDs will help to achieve the conversion of blue emission to the emission wavelength of the QDs. The schematic diagram demonstrating the color conversion process described above is shown in **Figure 4**. As it can be clearly seen from the figure, the process starts with the electrical injection of electrons and holes to the quantum wells of InGaN/GaN LEDs. Following their radiative and nonradiative recombination in the wells, carrier relaxation occurs. The radiative recombination results in the generation of photons with the wavelength corresponding to the bandgap of the InGaN wells (In incorporation defines the bandgap of the wells as stated above). The process of photon generation with this kind of electrical injection is called electroluminescence. These photons are absorbed by QDs placed on top of the device. The energy of the photons generated in the InGaN quantum wells should be higher than the bandgap of QDs to result in a successful excitation of charge carriers in QDs. Following the excitation of carriers with the incoming photons, the excited carriers recombine either radiatively or nonradiatively. Radiative recombination leads to the emission of photons with the wavelength corresponding to the bandgap of QDs. This kind of excitation of QDs which is a result of interaction between the photons of InGaN quantum wells and the charge carriers of QDs is called photoluminescence. **Figure 5** demonstrates the emission intensity of a color-converted InGaN/GaN LED incorporating QDs at 10, 20, 30, and 50 mA current levels.

The process of color conversion with the incorporation of semiconductor QDs is an equivalent method to the color conversion utilizing phosphors [12]. However, there are several drawbacks in using these phosphors for down-conversion. The bandwidth of emission in most

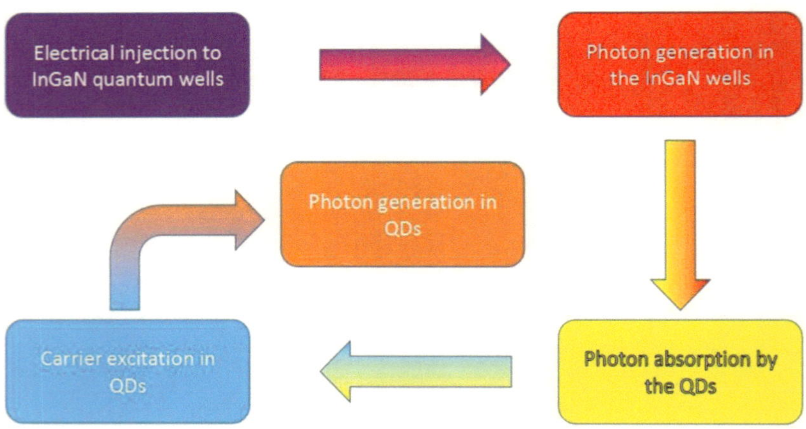

Figure 4. Schematic diagram color conversion in hybrid color-converted InGaN/GaN-QD LED.

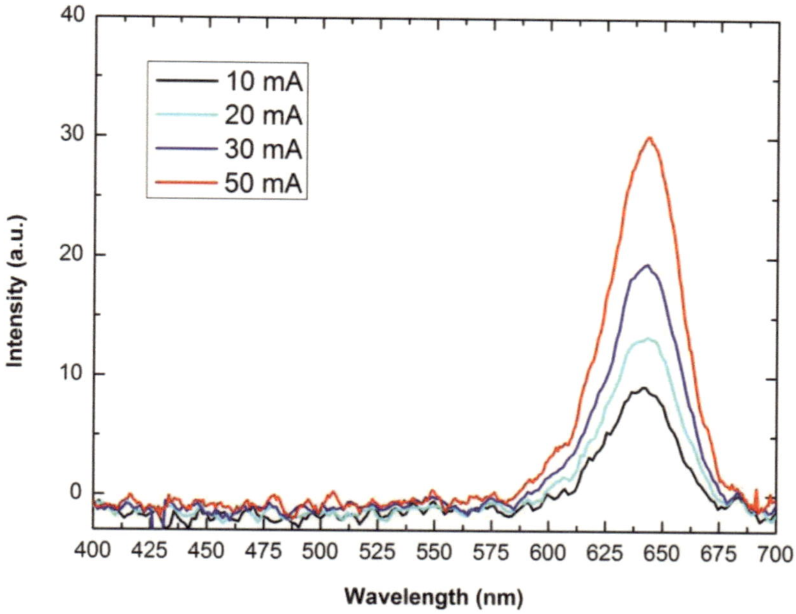

Figure 5. Emission intensity of a color-converted InGaN/GaN LED incorporating semiconductor nanocrystal QDs at 10, 20, 30, and 50 mA current levels.

phosphor compound materials is very large, almost spanning the whole visible spectrum. On the other hand, QDs emit with the FWHM of around 40 nm which is significantly smaller than that of the phosphors. Moreover, very large amount of phosphor material is required to achieve the full color conversion. In comparison, QDs can result in color conversion with significantly less amount of material. The optical absorption is another key parameter during the fabrication of a hybrid color-converted device. The color-converter materials should have decent absorption to exhibit high efficiency during the conversion process. QDs can absorb almost all the photons with the wavelength slightly shorter than the emission wavelength of these nanocrystals. On the other hand, phosphors can absorb only narrow range of wavelength. These superior properties of QDs over phosphors make them very promising candidates as efficient color-converter layers.

Although QDs are highly effective in color conversion, their localization in a suitable structure defines their efficiency. QDs generally exhibit significantly high quantum yield when dispersed in a medium-like toluene. However, making close-packed films out of these nanocrystals may result in a significant reduction of quantum yield. The main underlying reason for this behavior can be explained as follows. When QDs are dispersed in toluene, the separation distance between individual QDs is very large. This separation prevents any kind of close interaction between the nanocrystals. However, when they make close-packed solid films, the separation distance between the QDs is very short. This close construction gives rise to the interaction of QDs via nonradiative resonance energy transfer through dipole-dipole coupling process. When a close-packed film is excited with a source, the photogenerated electrons and holes in the QDs builds dipoles (this is called donor in the energy transfer process). These dipoles can create a mirror dipole in the QDs placed in sub-10 nm range (acceptor). The generation of the dipole in the adjacent QD is a nonradiative process owing to the absence of photon generation by the donor QD and absorption by the acceptor QD. Not all of the transferred dipole energies result in the radiative recombination in the acceptor QDs. Thus, a huge amount of energy is lost during the resonance energy transfer process. To prevent this energy transfer resulting from the close interaction of QDs, the QDs should be separated in their solid films. To realize this, QDs can be dispersed in a special matrix. This will help to reduce the quantum yield loss originating from the nonradiative energy transfer [13].

Another important change during the formation of solid films is the shift of emission wavelength. One of the main underlying reasons for this shift is the change in the medium. The refractive index strongly affects the emission wavelength. Moreover, the nonradiative resonance energy transfer between the nanocrystals also plays a significant role in the shift of the peak. The transfer mechanism is depicted in **Figure 6**. In very close proximity in their solid films, there is a high chance of energy transfer through nonradiative dipole-dipole coupling to occur. As it is well known, nanocrystals are not perfectly synthesized; there is a finite size distribution of the synthesized nanocrystals. Nanocrystals with smaller radius exhibit larger bandgap energy owing to the reverse proportionality of the bandgap energy with the nanocrystal radius in the calculation of quantum confinement. Thus, smaller nanocrystals (energy donor) tend to transfer their energy to larger nanocrystals (energy acceptor) close to them. Since the photoluminescence of the donor nanocrystals has a large spectral overlap with the absorbance of the acceptor nanocrystals as well (see **Figure 7**), the donors are able to transfer their excitons to the acceptors. These transferred excitons relax to the ground states and recombine for possible radiative emission. As a result of this transfer, collective emission intensity of the nanocrystals with smaller radius and higher energy (emitting with shorter wavelength) decreases. Moreover, emission intensity of the nanocrystals with larger radius

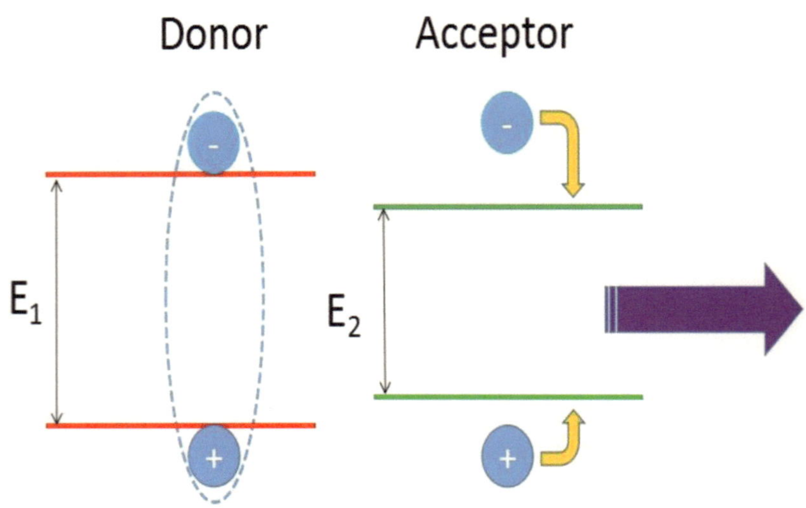

Figure 6. Energy transfer mechanism between smaller (donor) and larger (acceptor) nanocrystals.

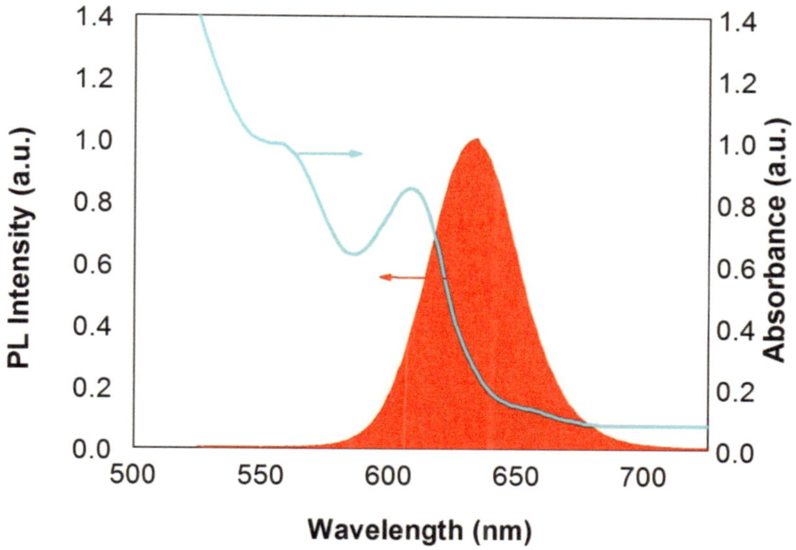

Figure 7. Spectral overlap between the photoluminescence and absorbance curves of nanocrystals.

and lower energy (emitting with longer wavelength) increases. This results in the red shift of the emission intensity when compared with their emission in toluene.

3. QD-incorporated hybrid white LEDs

As it is clearly discussed in the previous section, semiconductor QDs can effectively change emission color of a device by fully converting the incoming photons. On the other hand, it is possible to achieve a mixed color emission by utilizing the optimized amounts of the red, blue, yellow, and green QDs. In this context, **Figure 8** shows the CIE chromaticity diagram with the emission wavelengths and chromaticity coordinates. As it can be clearly seen from the diagram, white light is in the center, and it can be observed only by mixing several colors.

As it is well known, the main properties defining a white light emission of a high quality are its correlated color temperature (CCT), color rendering index (CRI), and luminous efficacy of optical radiation (LER). CRI measures how efficiently a white light emitting device reflects the real color of an illuminated object. In order to have a high-quality white light source, CRI should be higher than 90. LER is a measure of how well the produced light is perceived by the human eye. The unit of LER is lumens per watt. It is calculated with the following equation.

$$LER = 683 \quad \mathrm{lm}/\mathrm{W}_{\mathrm{op}} \frac{\int v(\lambda)s(\lambda)d\lambda}{\int s(\lambda)d\lambda} \tag{1}$$

$s(\lambda)$ is the spectral distribution of the radiated optical power and $V(\lambda)$ is the eye sensitivity function. Although mathematically the highest LER is 683 lm/W_{op}, it is almost impossible to achieve this number experimentally. A high-quality white light source should exhibit LER of

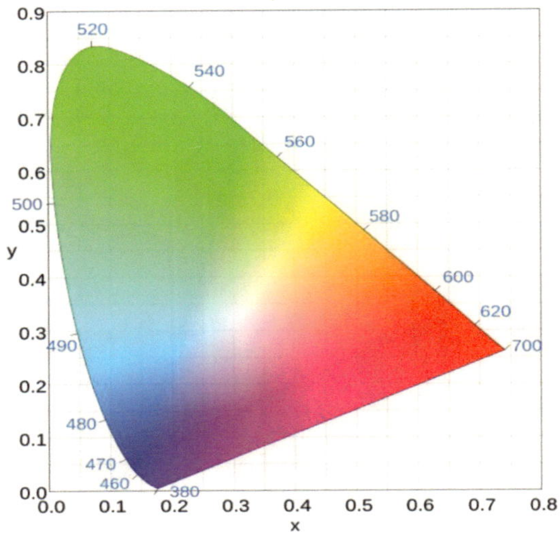

Figure 8. CIE Chromaticity diagram with emission wavelengths and chromaticity coordinates.

above 300 lm/W_{op} [14]. Another important photometric figure-of-merit is CCT. **Figure 9** demonstrates CCT chart on the chromaticity diagram to clearly understand the color quality difference between several CCT values. As it is seen from the figure, the amounts of individual colors define its chromaticity coordinates and consequently its CCT.

It indicates the temperature of a Planck black-body radiator whose perceived color most closely resembles that of the light-source. The optical output of a white light-emitting device can be either cool or warm white light. CCT of a warm white light source is below 3500 K. Warm white (right) and cool white light (left) emission are shown in **Figure 10**. Warm and cool white light sources differ in their areas of applications. For example, in the interior house

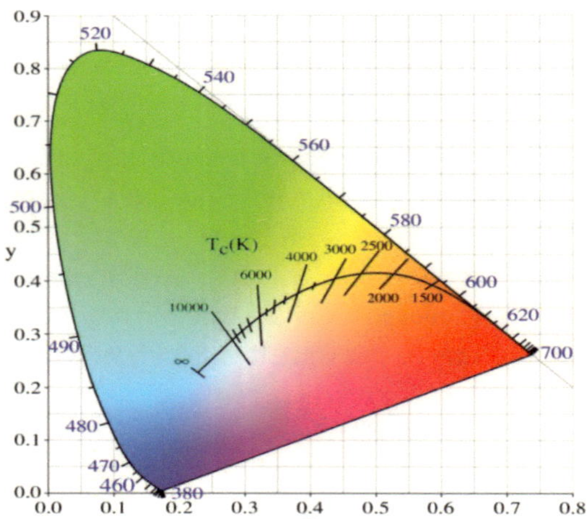

Figure 9. CIE chromaticity diagram with CCT chart.

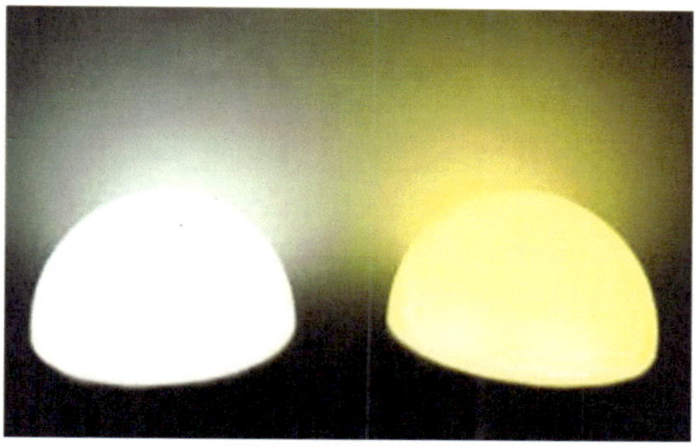

Figure 10. Cool (left) and warm (right) white light sources.

design, it is more suitable to use warm white light in the bedrooms, living rooms, and hallways, while cool white light is generally used in kitchen, study rooms, and bathrooms.

In general, to achieve a high-quality white LED with the incorporation of semiconductor QDs, the mixture of blue, green, yellow, and red emission is essential. If blue InGaN/GaN LED is used as an electrically injected device with nearly 450 nm emission, QDs with green, yellow, and red QDs are necessary to generate a white light with a high brightness. The schematics of the hybrid white LED utilizing blue InGaN/GaN LED is depicted in **Figure 11**. As it is shown in **Figure 11(a)** and **(b)**, hybrid white LEDs can be constructed by adding layered and blended QDs. In the layered architecture (**Figure 11(a)**), it is necessary for the QDs to be in the deposition order of red, yellow, and green that results in a device with highest efficiency.

Another architecture for generating white light is utilizing dual wavelength InGaN/GaN LEDs. In this design, only two kinds of semiconductor QDs are incorporated. Dual wavelength multiple quantum well InGaN/GaN LEDs are epitaxially grown on c-plane sapphire

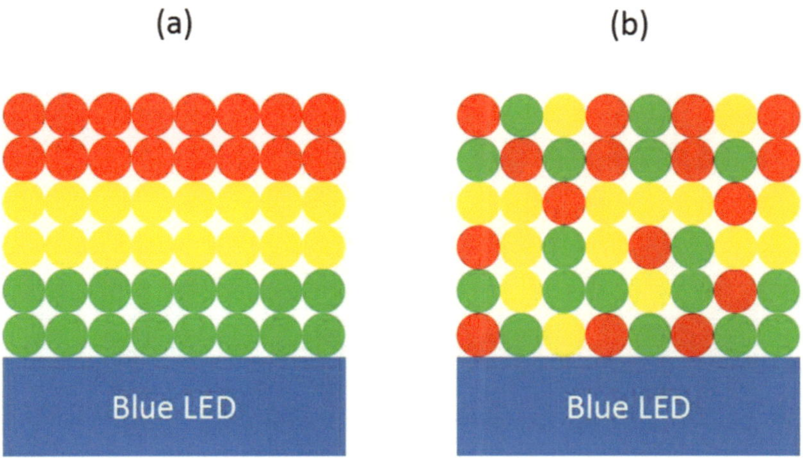

Figure 11. White LEDs constructed by adding (a) layered and (b) blended QDs on top of the blue InGaN/GaN LEDs.

substrates [15]. Unintentionally doped thick GaN layer (4 μm) is grown following the deposition of a thin low temperature (550°C) nucleation layer (30 nm). Then a 3 μm thick n-doped GaN layer is grown at high temperature. Si with a doping concentration of 5×10^{18} cm^{-3} was utilized as a p-type dopant. Three blue quantum wells (2.5 nm) were grown with GaN quantum barrier (10 nm) separation layers. An In composition of 15% was used to achieve blue emission from these three quantum wells. Subsequently, three quantum wells with higher In composition were grown to achieve green emission. p-Type doped 30 nm thick AlGaN layer was grown on top of a 10 nm GaN cap layer to serve as an electron-blocking layer. Utilizing the electron-blocking layer helps to prevent the leakage of excess electrons to the quantum wells to result in carrier imbalance. Finally, a 200 nm thick p-doped GaN layer was deposited on electron-blocking layer. Devices were fabricated with patterning, mesa etching, and electrode deposition. The device can emit the mixture of blue and green colors. The intensity of emission can be modified with the operation current of the device. In our architecture whose construction is described above, green quantum wells are closer to p-contacts when compared with blue quantum wells. Thus, at low current levels, green emission dominates the device output. However, once the current level is increased to a certain value, radiative recombination starts to happen more frequently in blue quantum wells as well. This will increase the blue emission intensity of the dual wavelength device. Once the device is fabricated, quantum dots with yellow and red (or amber) emission are placed on top of fabricated devices either in a layer or in a blended architecture. **Figure 12** shows the schematics of the dual wavelength multiple quantum well InGaN/GaN LED emitting blue and green colors covered with yellow and red semiconductor QDs.

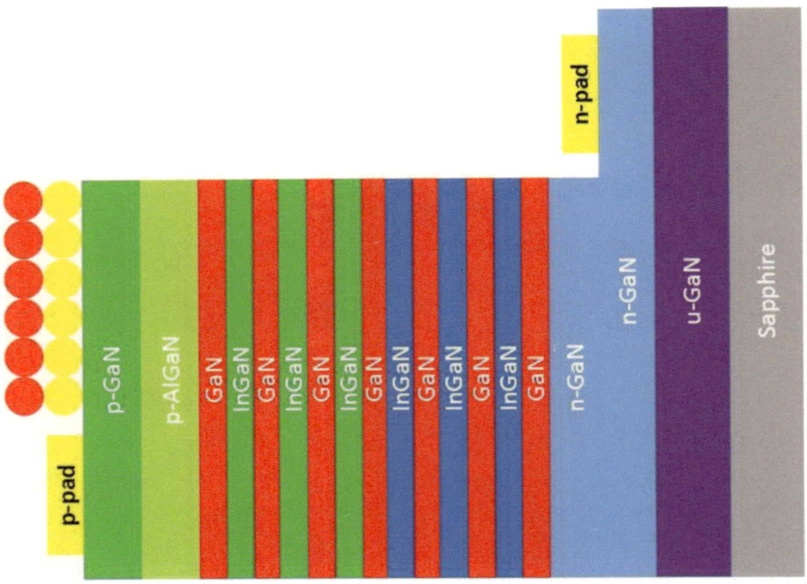

Figure 12. Dual wavelength multiple quantum well InGaN/GaN LED-emitting blue and green colors covered with yellow and red quantum dots.

4. FRET-enhanced hybrid LEDs

Forster resonance energy transfer (FRET) can enhance the optical power and power conversion efficiency of the conventional color converted and white LEDs. In this context, the relative quantum efficiency of color converter of QDs is increased by nonradiatively transferring extra excitons from the defect states of the closely placed donor QDs. The relative quantum efficiency enhancement mechanism of QDs is explained as follows. The optically excited semiconductor QDs contain excitons that recombine either radiatively or nonradiatively. The so-called "nonradiative" excitons are able to transfer their excitonic energy to the neighboring acceptor QDs before they recombine in defects in the host QDs. This increased the quantum efficiency of the acceptor QDs with increased emission yield. The process of transferring these excitons through nonradiative FRET process with dipole-dipole coupling is called exciton recycling [16].

Electronic band structure of FRET-converted LEDs is depicted in **Figure 13**. Excitons and/or charge carriers are transported to quantum wells following the electrical injection to blue LEDs. The radiative recombination in the InGaN (In composition of 15%) quantum wells leads to blue emission. This emission optically excites the donor semiconductor QDs. The excitons of these QDs are transferred to the acceptor QDs. The excited acceptor color-converter QDs emit with the emission wavelength corresponding to their bandgap energy. To support the existence of FRET process occurring between the donor and acceptor QDs, several types of experiments can be done. One of the most commonly known methods to examine FRET is measure the lifetimes of the acceptor and donor molecules with time-resolved fluorescence spectroscopy. If FRET occurs, donor's lifetime should be shortened owing to the exciton migration from these host molecules. On the other hand, the acceptor molecules should exhibit longer lifetime in the presence of their donor counterparts thanks to the exciton feeding. Another method to check whether exciton migration in donor-acceptor pairs is present or not is to acquire the photoluminescence excitation spectral behavior of the acceptor QDs in the presence and in the absence of the donor QDs. Although this method can also provide a strong argument on the presence/absence of FRET, the former time-resolved spectroscopy method is more effective in evaluating the efficiency of the FRET process.

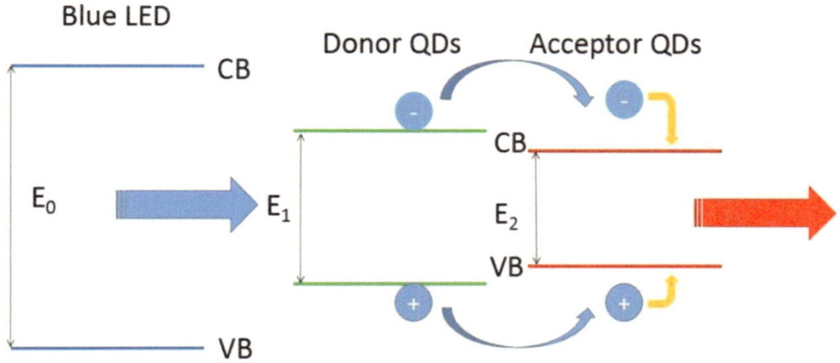

Figure 13. Electronic band structure of FRET-converted LEDs.

The FRET-based architecture above only demonstrates the enhancement process in the color-converted LED. However, it is possible to enhance the color quality of white LEDs with the utilization of FRET as well by individually increasing the quantum efficiency of less efficient QD components in the hybrid white emitting devices. Using this way, the intensity of the individual (green, yellow, or red) can be modified. This modification leads to the changes in the chromaticity coordinates of the white light and CCT.

5. Plasmon-enhanced hybrid LEDs

As it was stated above, FRET is a powerful concept to enhance the efficiency of the color-converter QDs incorporated in hybrid LEDs. Another method to increase the relative quantum yield of these semiconductor QDs is to make use of plasmon-exciton coupling mechanism [17]. Plasmon coupling can be realized by either forming localized surface plasmons or surface plasmon polaritons in the close vicinity of the emitter. Localized surface plasmons result in more pronounced absorption peaks. Moreover, their use within the device structures is more convenient owing to its simple configuration when compared with surface plasmon polaritons. The plasmonic absorption peaks of localized surface plasmons can be easily modified by changing the size of these particles. **Figure 14** shows the absorption spectra of Ag nanoparticles with different deposition thicknesses and same annealing condition. 10, 15, and 20 nm thick electron beam deposited and annealed Ag (films become nanoparticles following the deposition of such thin layers and high temperature annealing) exhibit 450, 504, and 666 nm absorption peaks, respectively.

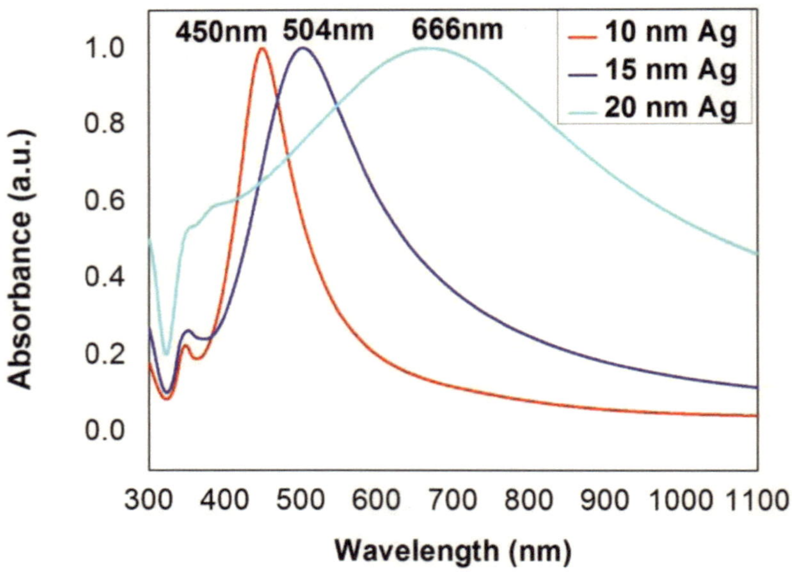

Figure 14. Absorption spectra of electron beam-deposited Ag layers with deposition thickness of 10, 15 and 20 nm.

http://dx.doi.org/10.5772/intechopen.68356

To achieve a successful enhancement in the emission yield, absorption spectrum of plasmonic metal structure should have a decent overlap with the luminescence spectrum of the emitter (semiconductor QD in this particular case). In this context, it is important to choose the correct metal material to achieve a good spectral overlap. Ag is a more convenient material for the emission in near UV and blue. On the other hand, Au can be utilized to get a plasmonic enhancement in the emitters with emission wavelength of more than 500 nm. Moreover, the position of plasmonic structure is also an important feature which needs extra care. The QD and the plasmonic metal should be in close proximity to realize an efficient coupling between them. On the other hand, if these two structures are placed very closely to each other, the emission of QDs will be strongly quenched thanks to nonradiative energy transfer from QDs to the adjacent metal structure. Thus, it is essential to optimize the relative locations of the QDs and the plasmonic metal structure to prevent the nonradiative energy transfer-induced emission loss and to achieve strong exciton-plasmon coupling simultaneously.

A thin metallic layer can be chemically grown on top of QDs to achieve plasmon-induced enhancement (**Figure 15**-left). However, placing a metal layer directly on the surface of QD would result in nonradiative resonance energy transfer-induced quenching of QD emission. Thus, it is important to insert a spacer shell layer in between the semiconductor QD and metallic shell. Moreover, the thickness of these shells (spacer and metal) cannot be too large; thick shells would block a significant amount of light coming out of QDs. Another useful mechanism to achieve strong coupling between the plasmons generated in metals and excitons in QDs is to make use of blended structure (**Figure 15**-right). In this configuration, chemically synthesized small metallic nanoparticles are mixed with QDs in solution. Later, they will make a solid blended film on a flat surface. However, direct contact of core QDs with metallic nanoparticle would again give rise to a strong quenching of emission owing to nonradiative energy transfer. To prevent nonradiative quenching, QDs can be synthesized in a core-shell configuration such as CdSe/ZnS with optimized shell thickness. In both of the configurations explained above, surface plasmons provide additional radiative channels for the excitons of QDs. This leads to the enhanced emission yield of QDs. The two powerful methods to examine the existence of plasmon-exciton coupling are photoluminescence measurement of QD films and time-resolved photoluminescence decay experiments. The former method shows the photoluminescence peak enhancement of QDs owing to the existence of plasmonic nanostructures in close vicinity. In most cases, there is a peak shift in the plasmon-incorporated films owing to the difference in the absorbance peak of metal and photoluminescence of the emitter; the spectral region of

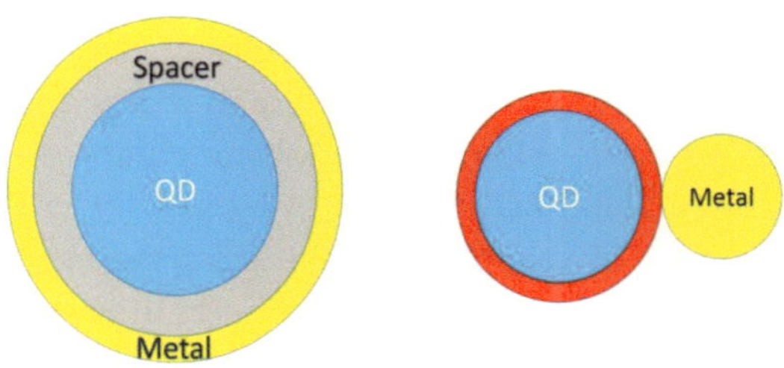

Figure 15. QD-plasmon coupling mechanisms with core-shell (left) and blended (right) configurations.

QDs corresponding to the absorbance peak location of metallic nanostructures gain maximum photoluminescence enhancement. This leads to the shift of photoluminescence peak toward the absorbance peak. The second useful method to examine the existence of plasmon-exciton coupling is to draw the photoluminescence decay curves of QD films in the presence and in the absence of the plasmonic nanostructures. Due to the presence of additional radiative channels in the QD-metal structure, the photoluminesce of this structure should decay faster; the reduction in the lifetime of QD film in the presence of plasmonic metal nanostructures is attributed to the strong plasmon-exciton coupling induced by the increased radiative recombination rate.

The optimized blended and core-shell configurations (**Figure 15**) can be incorporated as efficient color-converter materials on top of blue InGaN/GaN LEDs. These novel architectures would exhibit enhanced power conversion efficiency and optical power when compared with conventional color-converted hybrid LEDs utilizing pure QD film owing to the plasmon-induced quantum efficiency enhancement of QDs.

6. FRET-converted LEDs

Color conversion process in hybrid LED designs utilizing InGaN/GaN LED structures and QDs can be photonic or excitonic. In photonic color conversion, photons are generated in the quantum wells of InGaN/GaN LEDs following an efficient electrical injection, and they excite the color converter semiconductor QDs placed on top of the structure. In this design, there is a significant separation between QDs and the quantum wells of LEDs. On the other hand, excitonic color conversion process does not involve the generation of photons to excite the semiconductor QDs. In this configuration, excitonic energy of the InGaN quantum wells is directly transferred to the QDs through FRET (nonradiative dipole-dipole coupling). It is possible to achieve white light emission (or any other mixed color emission) by carefully controlling the emission intensity of InGaN quantum wells and the QD film.

The interaction of QDs and the quantum wells of InGaN/GaN LEDs can be realized through constructing several hybrid systems [18, 19]. One of the methods to achieve excitonic color conversion is to place QDs directly on top of the quantum wells (**Figure 16**-right). However, there should be a thin GaN cap layer with optimized thickness to control the amount of transferred nonradiative energy. Very small separation between the donor (InGaN quantum well) and the acceptor (QD) would result in higher emission intensity of QDs and lower intensity of InGaN wells. Thus, by modifying the thickness of cap layer, it is possible to optimize the quality of white light; CCT can be effective shifted. Moreover, QDs can be placed in between the nanopillars of InGaN/GaN LED structure to realize the interaction of QDs and sidewalls of InGaN quantum wells (**Figure 16**-left). Nanopillars can be fabricated either by etching the epitaxially grown bulk LED structure or by selectively growing LED structures in the holes of SiO2 layer on top of a sapphire substrate. Furthermore, LEDs with microholes can be fabricated and QDs can be inserted into these holes to observe the possible coupling between QDs and quantum wells. Energy transfer between the abovementioned donor and acceptor components are examined with photoluminescence, optical power measurements, and time-resolved photoluminescence spectroscopy studies.

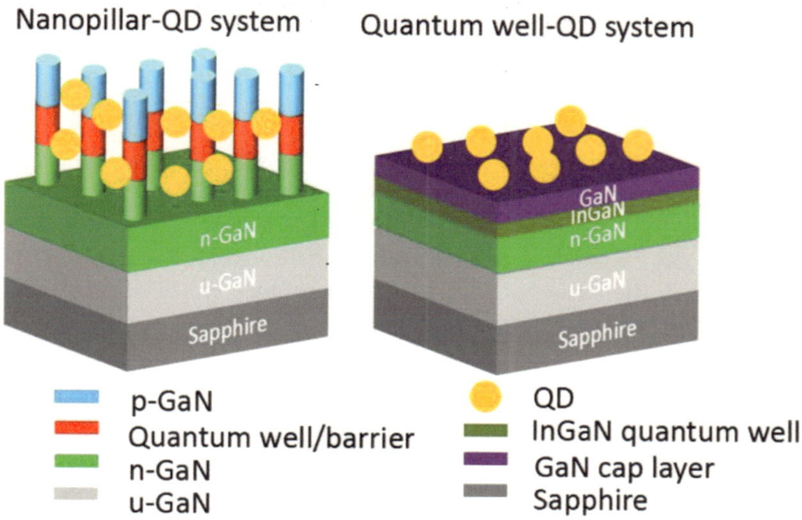

Figure 16. Schematics of nanopillar-QD (left) and quantum well-QD (right) systems.

Author details

Namig Hasanov

Address all correspondence to: namig@ntu.edu.sg

Division of Microelectronics, Electrical and Electronic Engineering, Nanyang Technological University, Singapore

References

[1] Smith AM, Nie S. Semiconductor nanocrystals: Structure, properties, and band gap engineering. Accounts of Chemical Research. 2010;**43**(2):190-200. DOI:10.1021/ar9001069

[2] Liu S, Liu W, Ji W, et al. Top-emitting quantum dots light-emitting devices employing microcontact printing with electric field-independent emission. Scientific Reports. 2016;**6**:22530. DOI:10.1038/srep22530

[3] Nakamura S, Mukai T, Senoh M. Candela-class high-brightness InGaN/AlGaN double-heterostructure blue-light-emitting diodes. Applied Physics Letters. 1994;**6**:1687-1689. DOI: http://dx.doi.org/10.1063/1.111832

[4] Akasaki I, Amano H, Koide Y, Hiramatsu K, Sawaki N. Effects of ain buffer layer on crystallographic structure and on electrical and optical properties of GaN and $Ga_{1-x}Al_xN$ ($0 < x \leqq 0.4$) films grown on sapphire substrate by MOVPE. Journal of Crystal Growth. 1989;**98**(1):209-219. DOI: http://dx.doi.org/10.1016/0022-0248(89)90200-5

[5] Amano H, Akasaki I, Hiramatsu K, Koide Y, Sawaki N. Effects of the buffer layer in metalorganic vapour phase epitaxy of GaN on sapphire substrate. Thin Solid Films. 1988;**163**:415-420. DOI: http://dx.doi.org/10.1016/0040-6090(88)90458-0

[6] Shiu GY. Chen KT, Fan FH, Huang KP, Hsu WJ, Dai JJ, Lai CF, Lin CF. InGaN light-emitting diodes with an embedded nanoporous GaN distributed Bragg reflectors. Scientific Reports. 2016;**6**:29138. DOI: 10.1038/srep29138

[7] Krishnamoorthy S, Akyol F, Rajan S. InGaN/GaN tunnel junctions for hole injection in GaN light emitting diodes. Applied Physics Letters. 2016;**105**:141104. DOI: http://dx.doi.org/10.1063/1.4897342

[8] Hsueh H, Ou S, Cheng C, Wuu D, Hong R. Performance of InGaN light-emitting diodes fabricated on patterned sapphire substrates with modified top-tip cone shapes. International Journal of Photoenergy. 2014;**2014**:7. DOI:10.1155/2014/796253

[9] Zhu L, Liu W, Zheng F, Gao Y, Liu B, Lu Y, Chen Z. Efficiency droop improvement in InGaN/GaN light-emitting diodes by graded-composition multiple quantum wells. IEEE Photonics Journal. 2013;**5**(2):8200208-8200208. DOI: 10.1109/JPHOT.2013.2245881

[10] Deng Z, Jiang Y, Ma Z, Wang W, Jia H, Zhou J, Chen H. A novel wavelength-adjusting method in InGaN-based light-emitting diodes. Scientific Reports. 2013;**3**:3389. DOI:10.1038/srep03389

[11] Huang C, Su Y, Chen Y, Wan C. Hybridization of CdSe/ZnS quantum dots on InGaN/GaN multiple quantum well light-emitting diodes for pink light emission. 2008. IEEE Photonics Global@Singapore, Singapore. 2008:1-3. DOI: 10.1109/IPGC.2008.4781324

[12] Sheu J, Chang S, Kuo C, Su Y, Wu L, Lin Y, Lai W, Tsai J, Chi G, Wu R. White-light emission from near UV InGaN-GaN LED chip precoated with blue/green/red phosphors. IEEE Photonics Technology Letters. 2003;**15**(1):18-20. DOI: 10.1109/LPT.2002.805852

[13] Xu L, Liu N, Xu J, Yang F, Ma Z, Chen K. Evolution of luminescence properties of CdTe quantum dots in liquid/solid environment. Journal of Nanoscience and Nanotechnology. 2011;**11**(11):9519-9522. DOI: https://doi.org/10.1166/jnn.2011.5282

[14] Phillips J, Coltrin M, Crawford M, Fischer A, Krames M, Mueller-Mach R, Mueller G, Ohno Y, Rohwer L, Simmons J, Tsao J. Research challenges to ultra-efficient inorganic solid-state lighting. Laser and Photonics Review. 2007;**1**(4):307-333. DOI: 10.1002/lpor.200710019

[15] Wen F, Huang L, Tong L, Huang D, Liu D. Luminescence properties of blue and green dual wavelength InGaN/GaN multi-quantum well light-emitting diode. Frontiers of Optoelectronics. 2009;**2**(4):446-449. DOI: 10.1007/s12200-009-0070-4

[16] Zhang X. Enhanced quantum efficiency in mixed donor-acceptor nanocrystal quantum dot monolayers. In: 2011 13th International Conference on Transparent Optical Networks, Stockholm; 2011. pp. 1-4. DOI: 10.1109/ICTON.2011.5971137

[17] Kim N, Hong S, Kang J, Myoung N, Yim S, Jung S, Lee K, Tu C, Park S. Localized surface plasmon-enhanced green quantum dot light-emitting diodes using gold nanoparticles. RSC Advances. 2015;**5**:19624-19629. DOI: 10.1039/C4RA15585H

[18] Xu X, Wang H. Resonant energy transfer between patterned InGaN/GaN quantum wells and CdSe/ZnS quantum dots. Nanoscale. 2016;**8**(1):342-347. DOI: 10.1039/C5NR05078B

[19] Achermann M, Petruska M, Kos S, Smith D, Koleske D, Klimov V. Energy-transfer pumping of semiconductor nanocrystals using an epitaxial quantum well. Nature. 2004; **429**:642-646. DOI: 10.1038/nature02571

Chapter 7

Valley Polarized Single Photon Source Based on Transition Metal Dichalcogenides Quantum Dots

Fanyao Qu, Alexandre Cavalheiro Dias, Antonio Luciano de Almeida Fonseca, Marco Cezar Barbosa Fernandes and Xiangmu Kong

Additional information is available at the end of the chapter

http://dx.doi.org/10.5772/intechopen.70300

Abstract

Photonic quantum computer, quantum communication, quantum metrology, and optical quantum information processing require a development of efficient solid-state single photon sources. However, it still remains a challenge. We report theoretical framework and experimental development on a novel kind of valley-polarized single-photon emitter (SPE) based on two-dimensional transition metal dichalcogenides (TMDCs) quantum dots. In order to reveal the principle of the SPE, we make a brief review on the electronic structure of the TMDCs and excitonic behavior in photoluminescence (PL) and in magneto-PL of these materials. We also discuss coupled spin and valley physics, valley-polarized optical absorption, and magneto-optical absorption in TMDC quantum dots. We demonstrate that the valley-polarization is robust against dot size and magnetic field, but optical transition energies show sizable size-effect. Three versatile models, including density functional theory, tight-binding and effective $k \cdot p$ method, have been adopted in our calculations and the corresponding results have been presented.

Keywords: single-photon source, quantum dots, transition metal dichalcogenides

1. Introduction

Traditional semiconductors have been used for decades for making all sorts of devices like diodes, transistors, light emitting diodes, and lasers [1]. Due to the advances of technology in fabrication, it is possible not only to make ever pure semiconductor crystals, but also to study heterostructures, in which carriers (electrons or holes) are confined in thin sheets, narrow lines, or even a point [1, 2]. Quantum dots (QDs) are zero-dimensional objects where all the three spatial dimensions are quantized with sizes smaller than some specific characteristic lengths, e.g., the exciton Bohr radius [1, 2]. Because of confinement, electrons in the QDs occupy discrete energy levels, in a similar way as they do in atoms [2, 3]. For these reasons, QDs are also referred to as artificial atoms [1, 2]. In spite of some similarities between the QD and the real atom, the former demonstrates several special characteristics. For instance, its size can vary from a few to hundreds of nanometers, and it can trap from a very small number of electrons (Ne < 10) to $50 - 100$ electrons or more [1]. In addition, the shape of the QD that can be tuned at will determine its spatial symmetry. In turn, the change in spatial symmetry modifies physical properties of the system. As known, the three-dimensional spherically symmetric QDs possess degenerate electron shells, 1*s*, 2*s*, 2*p*, 3*s*, 3*p*,... [3]. When the number of electrons is equal to 2, 10, 18, 36,..., the electron shells are completely filled [1], yielding a particularly stable configuration. In contrast, a two-dimensional cylindrically symmetric parabolic potential leads to formation of a two-dimensional shell structure with the magic numbers 2, 6, 12, 20,...[2, 4, 5]. Hence, the lower degree of symmetry in two-dimensional QDs leads to a smaller magic number sequence. Since the shape and size of the QDs can be precisely manufactured, the energy structure of the carriers in the QDs as well as their optical, transport, magnetic, and thermal properties can be engineered in a large scale [1–9].

Various techniques have been developed to produce the QDs such as etching, regrowth from quantum well structures, beam epitaxy, lithography, holograph patterning, chemical synthesis, etc [1, 2]. Consequently, many kinds of QDs emerge. According to the electrical property of their parent material, they can be classified into metal, semiconductor, or super-conducting dots. From geometry point of view, the QDs form two groups: two-dimensional [2, 4, 5, 8] or three-dimensional (3D) [1] dots. The former can be further divided into conventional 2D semiconductor QDs, such as self-assembled- and gated-QDs based on traditional semiconductor quantum wells [4–8, 10], and the novel QDs made from two-dimensional-layered materials (2DLMs) [11–15].

Atomically thin 2DLMs have revolutionized nanoscale materials science [16]. The interatomic interaction within layers is covalent in nature, while the layers are held together by weak van der Waals (vdW) forces. The family of 2D materials, which started with graphene [16], has expanded rapidly over the past few years and now includes insulators, semiconductors, semimetals, metals, and superconductors [17–19]. The most well studied 2D systems beyond graphene, are the silicene, germanene, stanine, and borophene, organic-inorganic hybrid perovskites, insulator hexagonal boron-nitride [17, 18], the anisotropic semiconductor phosphorene, transition metal-carbides, -nitrides, -oxides, and -halides, as well as the transition metal dichalcogenides (TMDCs) [20–28]. Compared with traditional semiconducting materials, the 2DLMs take advantage of inherent flexibility and an atomically-thin geometry. Moreover, because of their free dangling bonds at interfaces [25, 29, 30], two-dimensional-layered materials can easily be integrated with various substrates [17]. They can also be fabricated in complex-sandwiched structures or even suspended to avoid the influence of the substrate [31]. The monolayer TMDCs with infinite geometry exhibit strong carrier confinement in one dimension but preserve the bulk-like dispersion in the 2D plane. In contrast, electrons in a TMDC QD are restricted in three dimensions, which present size tunable electronic and optical properties in addition to the remarkable characteristics related to spin-valley degree of freedom inherited from its 2D bulk materials. Very recently, graphene QDs (GQDs) have attracted intensive research interest due to their high transparency and high surface area. Many remarkable applications ranging from energy conversion to display to biomedicine are prospected [11]. Nevertheless, from quantum nano-devices point of view, the TMDCs have advantages over graphene. For instance, the semiconducting TMDCs have a band gap large enough to form a QD using the electric field, as shown in **Figure 1**, unlike etched GQDs made on semi-metallic graphene.

The applications of quantum dots are still mostly restricted to research laboratories, but they are remarkable due to the fact that QDs provide access to the quantum mechanical degrees of freedom of few carriers. Single electron transistors [1–6], the manipulation of one [4–7] or two [1–3] electron spins, manipulation of a single spin in a single magnetic ion-doped QD [4–7] are only some examples. Optically active quantum dots can also be used in both quantum communication and quantum computation [4–7, 12–15]. The emerging field of quantum information technology, as unconditionally quantum cryptography, quantum-photonic communication and computation, needs the development of individual photon sources [12–15, 32, 33]. Recently, individual photon emitters based on defects in TMDC monolayers with different sample types (WSe_2 and $MoSe_2$) have been reported, but only operate at cryogenic temperatures [12, 13, 32–35]. In addition, it should be kept in mind that the presence of defects is not always beneficial for the PL signal. For instance, defect-mediated nonradiative recombination might result in an internal quantum yield droop in the defective TMDCs [26–28]. In this context, single quantum emitter based on QD is desirable [14, 15].

In this chapter, we show the optical and magneto-optical properties of the TMDC QD's. We choose MoS_2, which has been widely studied in the literature as our example. Three versatile models including density functional theory, tight-binding, and effective $k \cdot p$ approach have been adopted in our calculations and the corresponding results have been presented. We show that the valley-polarization is robust against dot size and magnetic field, but the optical transition energies show sizable size effect. Based on the computed optical absorption spectra, a novel kind of valley-polarized single-photon source based on TMDC quantum dot is proposed [15].

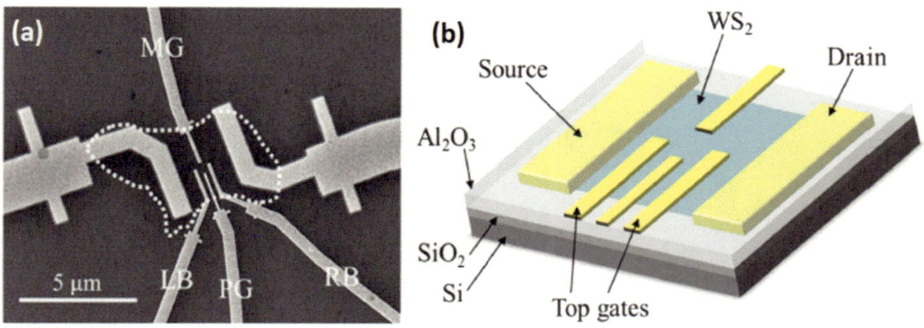

Figure 1. (a) Scanning electron microscope image of the WS_2 quantum dot studied. The WS_2 flake is highlighted by the white dotted line, and the four top gates are labeled as MG, LB, PG, RB. The scale bar represents 5 μ m. (b) Three-dimensional schematic view of the device [20]. (Copyright 2015 by the Royal Society of Chemistry. Reprinted with permission).

2. Physical properties of transition metal dichalcogenides

2.1. Electronic band structure of transition metal dichalcogenides

Layered TMDCs have the generic formula MX_2, where M stands for a metal and X represents a chalcogen. The monolayer crystal structure of the MoS_2, which is one of the most studied TMDCs in the literature is shown in **Figure 2**. Notice that the monolayer MoS_2 has a trigonal prism crystal structure. An inversion asymmetry results in a large direct gap semiconductor with the gap lying at the two inequivalent *K*-points of the hexagonal Brillouin zone.

The major orbital contribution at the edge of the conduction band (CB) is from $d_{3z^2-r^2}$ orbital of the metal, plus minor contributions from p_x, and p_y orbitals of chalcogens. On the other hand, at the edge of the valence band (VB) at the *K*-point, the most important orbital contribution is due to a combination of d_{xy} and $d_{x^2-y^2}$ of the metal, which hybridize to p_x and p_y orbitals of the chalcogen atoms [24]. In addition, there is a strong spin-orbit interaction (SOI), especially in the valence band. The spatial inversion asymmetry along with strong SOI leads to a spin-valley coupling, which results in the spin of electron being locked to the valley [20, 21, 36], as demonstrated in **Figure 3(a)**. Furthermore, the band structure in the *K*-valley presents time reversal symmetry with that in *K'*-valley. And, it changes dramatically as the number of the layers of the TMDC increases. When the thickness increases, the band gap in MoS_2 and other group VI TMDCs decreases, and more importantly, the material becomes an indirect gap semiconductor [37], as shown in **Figure 3(b)**.

In order to get insight into the physical origins of the band gap variations with the number of layers, **Figure 4** shows evolutions of the band gaps (a) and band edges (b) of MoS_2 as a function of the number of layers. Notice that with increasing layer thickness, the indirect band gap (Γ–K, Γ–Λ) becomes smaller, while the direct excitonic transition (K–K) only slightly changes, as illustrated in **Figure 4(a)**. Note also that the monolayer (n = 1) MoS_2 is a direct band gap semiconductor, but it becomes an indirect band gap semiconductor when the number of layers is larger than one. This phase transition is also clearly demonstrated in **Figure 4(b)**. In addition, as the number of layers increases, both the VB and CB edges at *K*-valley exhibit only slight variation while the degeneracy at Γ is lifted and a splitting of the bands occurs, which

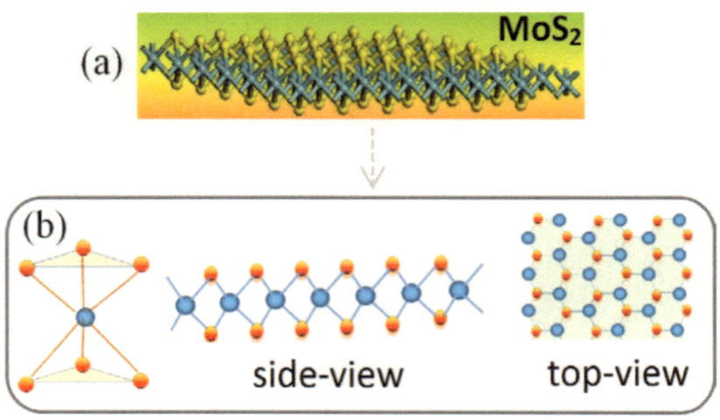

Figure 2. (a) Schematic diagram of crystal structure of MoS_2. (b) Coordination environment of Mo (blue sphere) in the structure (the left panel); the middle and the right panels correspond to side- and top- views of the monolayer MoS_2 lattice. Sulfur is shown as golden spheres.

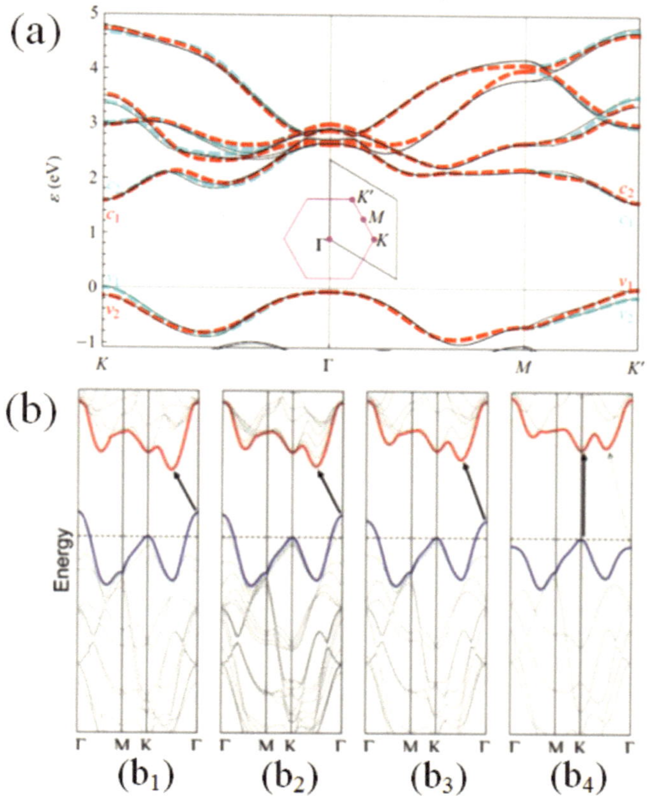

Figure 3. (a) Band structure of monolayer MoS_2. The solid curves were obtained using the QUANTUM ESPRESSO package with fully relativistic pseudopotentials under the Perdew-Burke-Ernzerhof generalized-gradient approximation, and a $16 \times 16 \times 1$ k grid. The dashed curves were calculated from the tight-binding model, with cyan (red) representing states that are even (odd) under mirror operation with respect to the Mo plane. $v_{1,2}$ and $c_{1,2}$ label the bands close to the valence and conduction band edges near the K- and K'-points. The inset shows the hexagonal Brillouin zone (pink) associated with the triangular Bravais lattice of MoS_2 and an alternate rhombohedral primitive zone (black), and labels the principle high-symmetry points in reciprocal space [36] (Copyright 2015 by American Physical Society, Reprinted with permission). (b) Calculated band structures of (b1) bulk MoS_2, (b2) quadrilayer MoS_2, (b3) bilayer MoS_2, and (b4) monolayer MoS_2. The solid arrows indicate the lowest energy transitions. Bulk MoS_2 is characterized by an indirect bandgap. The direct excitonic transition occurs at K point with a higher transition energy than that of indirect one [30]. (Copyright 2010 by the American Chemical Society. Reprinted with permission).

pushes the VB maximum to higher energy. The variation of the band gap is largely driven by the variation of the VB at Γ point. Going from a monolayer to a bilayer significantly raises the VB at Γ, resulting in a transition from the direct K–K gap to an indirect Γ–K gap. This dramatic change of electronic structure in monolayer MoS_2 results in the jump in monolayer photoluminescence efficiency.

2.1.1. *Massive Dirac fermions*

To gain insight of physics around the K- and K'-points, one can reduce multi-band tight-binding model to a two band **k.p** model, using Löwdin partitioning method [38]. For the monolayer TMDCs, one gets the Hamiltonian in the first order of k approximation,

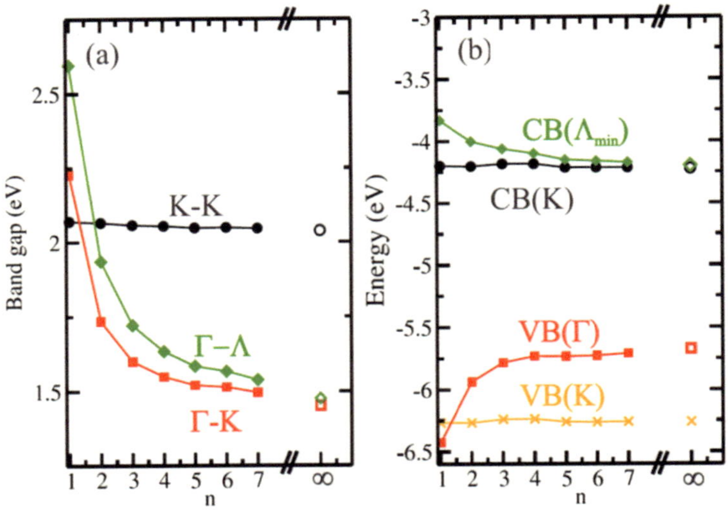

Figure 4. (a) Evolution of the band gaps as a function of the number of layers (n). The black circles ($K-K$), red squares (Γ–K), and green diamonds (Γ–Λ) indicate the magnitude of the different band gaps. Hollow symbols indicate the bulk band gaps. (b) Position of the band edge with respect to the vacuum level for the VB at the K-point (orange crosses), VB at Γ(red squares), CB at the K-point (black circles), and CB at Λ (green diamonds) [29]. (Copyright 2014 by the American Physical Society. Reprinted with permission).

$$H(\mathbf{k}, \tau, s) = \begin{pmatrix} \Delta/2 & at(\tau k_x - ik_y) \\ at(\tau k_x + ik_y) & -\Delta/2 + \tau s\lambda \end{pmatrix} \quad (1)$$

where $\Delta = 1.6\ eV$ is the band gap, $t = 1.1\ eV$ is the effective hopping parameter, $\lambda = 0.075\ eV$ is the SOI parameter, $a = 3.19\ \mathring{A}$ is the lattice parameter [39], $\tau = \pm 1$ is the valley index. The eigenvalues and eigenvectors can be derived straightforwardly as follows:

$$E_{\pm}(k_x,\ k_y) = \frac{\tau\lambda s}{2} \pm \sqrt{\frac{(\Delta - \lambda\tau s)^2}{4} + t^2a^2(k_x^2 + k_y^2)} \quad (2)$$

and

$$\begin{aligned} |c,\ \vec{k}\ \tau,\ s_z\rangle &= |s_z\rangle \otimes \begin{pmatrix} cos\left(\frac{\vartheta_n}{2}\right) \\ \tau sin\left(\frac{\vartheta_n}{2}\right) e^{i\tau\phi_{\vec{k}}} \end{pmatrix} \\ |v,\ \vec{k}\ \tau,\ s_z\rangle &= |s_z\rangle \otimes \begin{pmatrix} -\tau sin\left(\frac{\vartheta_n}{2}\right) e^{-i\tau\phi_{\vec{k}}} \\ cos\left(\frac{\vartheta_n}{2}\right) \end{pmatrix} \end{aligned} \quad (3)$$

$$cos\vartheta_n = \frac{\Delta + (-1)^n\lambda_{so}}{2\sqrt{(\Delta + (-1)^n\lambda_{so})^2 + 4t^2a^2k^2}} \quad (4)$$

$$tan(\phi_{\vec{k}}) = \frac{k_y}{k_x} \tag{5}$$

The energy dispersion around the *K*- and *K′*- points, described by Eq. 2, is shown in **Figure 5**. In order to see the reliability of the $\boldsymbol{k} \cdot \boldsymbol{p}$ approach, **Figure 6** plots the energy spectrum of

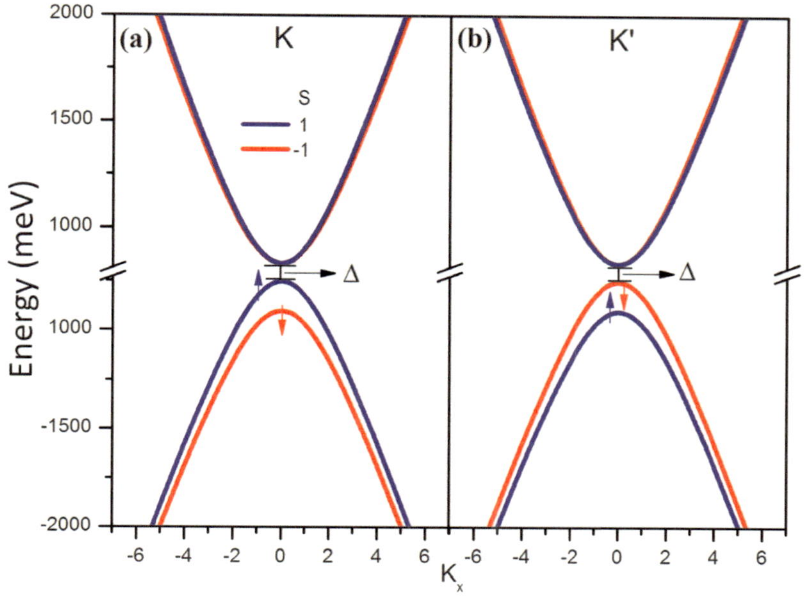

Figure 5. Energy dispersion in (a) *K*-valley and (b) *K*′-valley obtained by the $\mathbf{k} \cdot \mathbf{p}$ model.

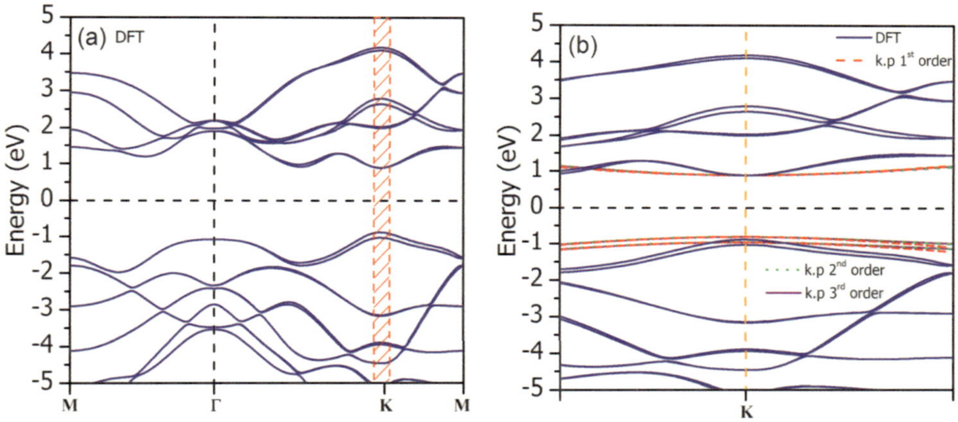

Figure 6. (a) Quasiparticle band structure of monolayer MoS_2 calculated by the density functional theory (DFT). (b) A blowup of the rectangular red area in (a). The blue (solid), red (dashed), green (dotted), and purple (dot-dashed) curves correspond to the results obtained by the DFT, and $\mathbf{k} \cdot \mathbf{p}$ theory of the first order, second order, and third order, respectively. Notice that around the *K*-point, all methods give almost identical results.

monolayer MoS_2 calculated by the first principle and $\boldsymbol{k} \cdot \boldsymbol{p}$ model. It can be found that in the vicinity of the K–(K')–point, they have very good agreement.

2.1.2. *Landau levels of monolayer MoS_2*

For a perpendicular magnetic field applied to the MoS_2 sheet, we use Peierls substitution $K_i \to \Pi_i = K_i + (e/\hbar)A_i$, where e is the elementary charge. In the Landau Gauge $\vec{A} = (0, Bx)$, we define the operators $\Pi_\pm = \tau\Pi_x \pm i\Pi_y$, which have the following properties:

$$[\Pi_-, \Pi_+] = (2\tau/l_B^2) \tag{6}$$

where the magnetic length is $l_B = \sqrt{\hbar/eB} \approx \left(25.6/\sqrt{B}\right)$. Using these operators, the destruction and creation operators are introduced in the following way:

$$\begin{aligned} \hat{b} &= \left(l_B/\sqrt{2}\right)\Pi_-, \\ \hat{b}^\dagger &= \left(l_B/\sqrt{2}\right)\Pi_+ \end{aligned} \tag{7}$$

in the K-($\tau = 1$) valley, and

$$\begin{aligned} \hat{b} &= \left(l_B/\sqrt{2}\right)\Pi_+, \\ \hat{b}^\dagger &= \left(l_B/\sqrt{2}\right)\Pi_-. \end{aligned} \tag{8}$$

in the K' ($\tau = -1$) valley. Then, the Hamiltonian in the presence of a perpendicular magnetic field can be well described by

$$\mathrm{H}^{\tau=1} = \begin{pmatrix} \frac{\Delta}{2} & ta\left(\sqrt{2}/l_B\right)\hat{b} \\ ta\left(\sqrt{2}/l_B\right)\hat{b}^\dagger & -\frac{\Delta}{2} + s\lambda \end{pmatrix}, \tag{9}$$

$$\mathrm{H}^{\tau=-1} = \begin{pmatrix} \frac{\Delta}{2} & ta\left(\sqrt{2}/l_B\right)\hat{b}^\dagger \\ ta\left(\sqrt{2}/l_B\right)\hat{b} & -\frac{\Delta}{2} - s\lambda \end{pmatrix}. \tag{10}$$

It is worth to mention that since the Zeeman effect is vanishingly small ($< 5\,meV$), it is neglected. After some algebra calculations, the Landau levels are obtained as follows:

$$E_\pm(\omega_c, n) = \frac{\lambda\tau s}{2} \pm \sqrt{\frac{(\Delta - \lambda\tau s)^2}{4} + t^2a^2\omega_c^2 n} \tag{11}$$

where n is an integer and $n \geq 1$, $\omega_c = \sqrt{2}/l_B$. The corresponding Landau fan diagrams of the monolayer MoS_2 are shown in **Figure 7**. The corresponding eigenfunctions are given by

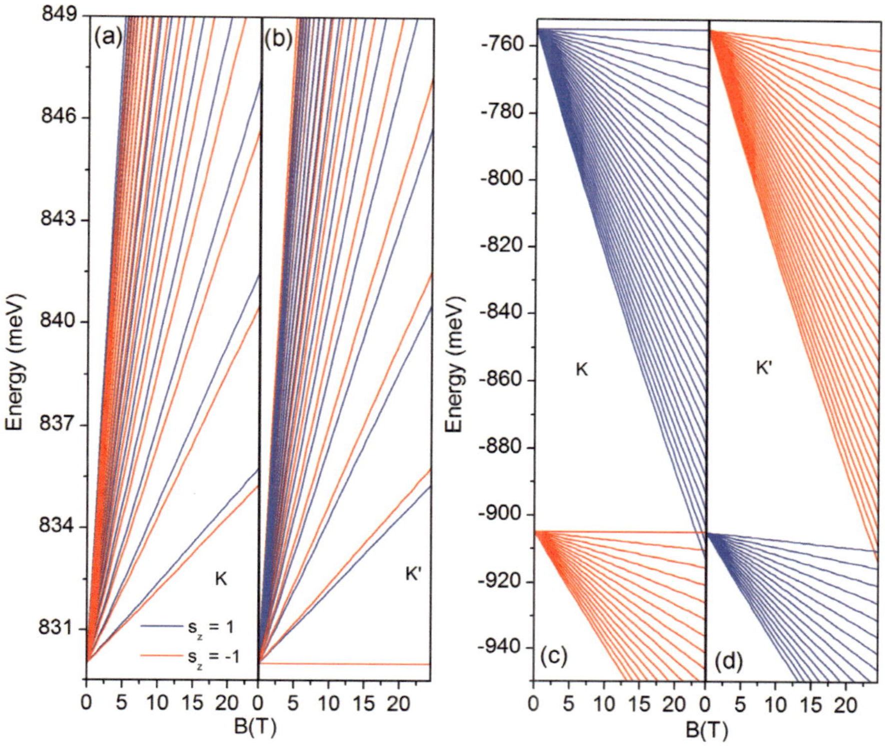

Figure 7. Conduction band Landau levels of monolayer MoS_2 in the vicinity of *K*- (a) and *K*′- (b) valleys with spin-orbit interaction (SOI). (c) and (d) are the same as (a) and (b), but for the valence band. The blue and red lines correspond to the spin-up and spin-down Landau levels, respectively.

$$\mathbf{\Psi}_{n,\pm}^{\tau=1} = \frac{1}{N_{\tau=1}^{n}} \begin{pmatrix} -\alpha_{\lambda,s,\pm}^{n}\phi_{n-1} \\ \phi_{n} \end{pmatrix} \tag{12}$$

and

$$\mathbf{\Psi}_{n,\pm}^{\tau=-1} = \frac{1}{N_{\tau=-1}^{n}} \begin{pmatrix} \phi_{n} \\ -\beta_{\lambda,s,\pm}^{n}\phi_{n-1} \end{pmatrix}, \tag{13}$$

where

$$\alpha_{\lambda,s,\pm}^{n} = \frac{ta\left(\sqrt{2}/l_B\right)\sqrt{n}}{\Delta/2 - E_{\pm}}, \tag{14}$$

$$\beta_{\lambda,s,\pm}^{n} = \frac{ta\left(\sqrt{2}/l_B\right)\sqrt{n}}{(-\Delta/2 - \lambda s) - E_{\pm}}, \tag{15}$$

$$N_{\tau=1}^{n,\pm} = \sqrt{(\alpha_{\lambda_{so},s_z,\pm}^{n})^2 + 1}, \tag{16}$$

$$N^{n,\pm}_{\tau=-1} = \sqrt{(\beta^{n}_{\lambda_{so},s_z,\pm})^2 + 1}, \tag{17}$$

$$\phi_n = \sqrt{\frac{1}{2^n n!}}\left(\frac{m\omega}{\pi\hbar}\right)^{1/4} e^{-m\omega r^2/2\hbar} H_n\left(\sqrt{\frac{m\omega}{\hbar}}r\right), \tag{18}$$

$$H_n(x) = (-1)^n e^{x^2}\frac{d^n}{dx^n}e^{-x^2}. \tag{19}$$

The eigenfunctions can be written in a compact form as,

$$\Psi^{\tau}_{n,\pm} = \frac{1}{N^{n,\pm}_{\tau}}\begin{pmatrix} c^{\tau,\pm}_{1,n}\phi_{n-(\frac{\tau+1}{2})} \\ c^{\tau}_{2,n,\pm}\phi_{n-(\frac{1-\tau}{2})} \end{pmatrix}. \tag{20}$$

For the special case in which $n = 0$, the eigenvalues become

$$E^{\tau=1}_{n=0} = -\frac{\Delta}{2} + \lambda_{so}s_z, \tag{21}$$

$$E^{\tau=-1}_{n=0} = \frac{\Delta}{2} \tag{22}$$

and corresponding eigenfunctions turn out to be

$$\Psi^{\tau=1}_{0} = \begin{pmatrix} 0 \\ \phi_0 \end{pmatrix}, \tag{23}$$

$$\Psi^{\tau=-1}_{0} = \begin{pmatrix} \phi_0 \\ 0 \end{pmatrix}. \tag{24}$$

2.2. Optical selection rules

In monolayer TMDCs, both the top of valence bands and the bottom of conduction bands are constructed primarily by the *d*-orbits of the transition metal atoms. The giant spin-orbit coupling splits the valence bands around the K (K') valley by 0.5 eV, for MoS_2 while the conduction band splitting is neglectable. In addition, time reversal symmetry (TRS) leads to the opposite spin splitting at the K- and K'-valleys. Namely the Kramers doublet (K,↑) and (K',↓) are separated from the other doublet (K',↑) and (K, ↓) by the spin–orbit interaction (SOI) splitting, as shown in **Figure 8**.

We assume that the monolayer TMDCs are exposed to light fields with the energy $\hbar\omega$ and wave vector k_l, which is orthogonal to the monolayer plane and much smaller than $1/a$. Up to the first order approximation, the light-matter interaction Hamiltonian is described by,

$$\mathcal{H}_{L-M} = e\mathbf{v}\cdot\mathbf{A}_{in}, \tag{25}$$

with the light field $\mathbf{A}_{in} = A_0\hat{\alpha}\cos(\mathbf{k}_l\cdot\mathbf{r} - \omega t)$ and $\mathbf{v} = (1/\hbar)\nabla_{\mathbf{k}}\mathcal{H}$ being propagation velocity [15]. Here, A_0 and $\hat{\alpha}$ stand for the amplitude and orientation of the polarization field,

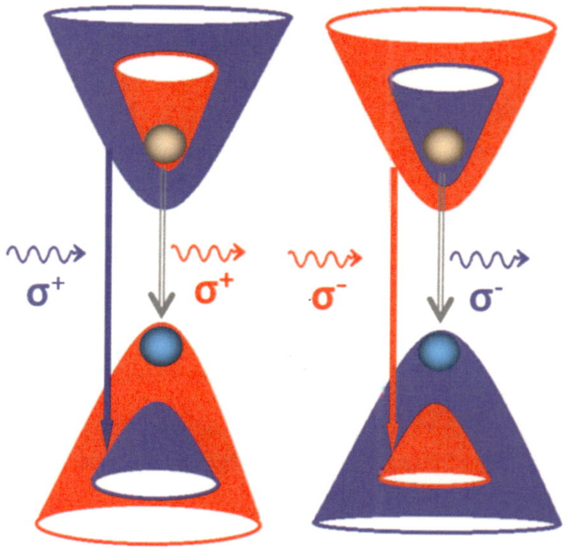

Figure 8. Schematic illustration of optical transition rules of the valley and spin in the K- (the left) and K'- (the right) valleys. The red (blue) color represents spin-up (down) states.

respectively. For an electron being excited by an incident photon from its initial state $|i\rangle$ to a final state $|f\rangle$, the transition probability is given by the Fermi's golden rule as, $W_{fi} = 2\pi/\hbar|\langle f|\mathcal{H}_{L-M}|i\rangle|^2 n(E)$, where $n(E)$ is the density of states available for the final state. Since the absorption intensity I is proportional to the transition rate, it can be evaluated by,

$$I = \sum_{m_c, m_v, n_c, n_v} | < \mathbf{\Psi}_c|\mathcal{H}_{L-M}|\mathbf{\Psi}_v> |^2 \Lambda \Upsilon, \tag{26}$$

where $\mathbf{\Psi}_c$ ($\mathbf{\Psi}_v$) is the conduction (valence) band wave function, $\Lambda = \gamma/\pi\{[\omega - (E_c(m_c, n_c) - E_v(m_v, n_v))]^2 + \gamma^2\}$, and $\Upsilon = f_c - f_v$, with f_i Fermi-Dirac distribution function, n_i the principal quantum number, m_i the quantum number associating with orbital angular momentum, $i = c$, and v referring to conduction and valence band, respectively, $m_c = m$ and $m_v = m'$, γ a parameter determined by the Lorentzian distribution.

For a circularly polarized (CP) light, $\hat{\alpha} = (1, \cos(\omega t - \sigma\pi/2), 0)^{\mathrm{T}}$, with $\sigma = \pm 1$ denoting the corresponding positive and negative helicities and T stands for the transpose of a matrix, then the perturbed Hamiltonian becomes,

$$\mathcal{H}_{L-M} = e^{i\omega t}\hat{W}_\sigma{}^{+} + e^{-i\omega t}\hat{W}_\sigma, \tag{27}$$

with $\hat{W}_\sigma = etaA_0(\tau\sigma_x + i\sigma\sigma_y)/2\hbar$. For example, for $\tau = 1$, $s_z = \pm 1$, the transition rate for the monolayer TMDCs is determined by,

$$\langle c, \vec{k}\,|\mathrm{H}_{L-M}^{CP,+}|v, \vec{k}\,\rangle = \frac{2m_0 at}{\hbar}\,\delta_{s_{zv}, s_{zc}}(sin\phi_{\vec{k}} + cos\phi_{\vec{k}} cos\vartheta_n) \tag{28}$$

and

$$\langle c, \vec{k}\, |\mathrm{H}_{L-M}^{CP,-}| v, \vec{k}\, \rangle = 0 \tag{29}$$

The optical transition rate for $\tau = -1$ and $s_z = \pm 1$ can be obtained by replacing + with −in the H_{L-M}. From Eqs. (29) and (30), notice that under CP light excitation, a valley and spin polarized emission or absorption light is expected in monolayer MoS_2, as shown in **Figure 8**. In contrast, linearly polarized light does not present valley-selected emission and absorption spectra because both *K*- and *K′*- valleys absorb light simultaneously.

2.3. Valley polarized photoluminescence and excitonic effects of the monolayer TMDCs

In monolayer TMDCs, strong Coulomb interactions due to reduced screening and strong 2D confinement lead to exceptionally high binding energies for excitons [23, 24, 36], which allow them be able to survive even at room temperature. Hence, the typical absorption spectra are usually characterized by strong excitonic peaks marked by A and B, located at 670 and 627 nm, respectively. The strong spin-orbit interaction in the valence band gives rise to a separation between them, as shown in **Figure 9**. In addition, an injection of electrons into the conduction band of MoS_2, which can be realized by gate-doping [26], photoionization of impurities [28], substrates [25] or functionalization layers [22, 27], leads to the formation of negatively charged excitons (X^-). The peak of the X^- is positioned at a lower energy side of neutral exciton with a binding energy about 36 meV for MoS_2, see the peak indicated by X^- in **Figure 10**. In addition, the emergence of the charged exciton is accompanied by a transfer of spectral weight from the exciton. Therefore, the intensity ratio between a neutral and charged exciton can be tuned externally. Besides, with increasing the nonequilibrium excess electron density, a red-shift of the excitonic ground-state absorption due to Coulomb-induced band gap shrinkage occurs. It is also worth to point out that on the one hand the trion can provide a novel channel for exciton relaxation, and on the other hand, it can also be excited by an optical phonon into an excitonic state to realize an upconversion process in monolayer WSe_2.

In the regime of high exciton density, the exciton-exciton collision leads to exciton annihilation through Auger process or formation of biexciton in the monolayer TMDCs. The biexciton is identified as a sharply defined state in the PL, see P_0 in **Figure 10** and also *XX*-peak in **Figure 11**. The nature of the biexcitonic state is supported by the dependence of its PL intensity on the excitation laser power. At low excitation laser intensity, the peaks P_0 and *X* grow superlinearly and linearly with incident laser power, whereas they increase sub-quadratically and sublinearly with the laser power at sufficiently high laser fluence. The large circular polarization of P_0 emission provides a further support for this assignment.

The polarization of the photoluminescence from the TMDCs, which is defined by $\eta = (I_{\sigma^+} - I_{\sigma^-})/(I_{\sigma^+} + I_{\sigma^-})$, inherits that of the excitation source, where I_{σ^+} (I_{σ^-}) is PL intensity of right (left) hand circularly polarized light. **Figure 11** illustrates photoluminescence spectra of monolayer WSe_2 excited by near-resonant circularly polarized radiation at 15 K. Notice that the peaks for *X*, X^-, and *XX* all exhibit significant circular polarization. In addition to biexciton emission, the *X* and X^- emission bands also exhibit strong valley polarization.

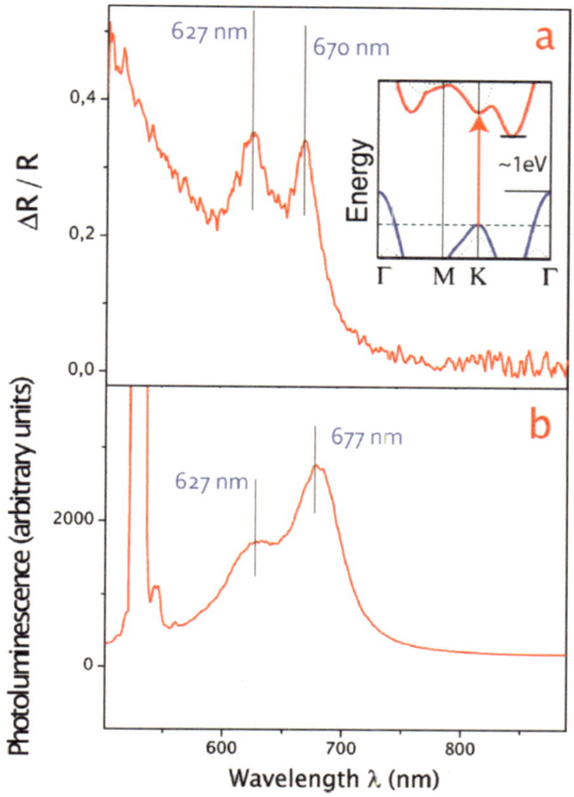

Figure 9. Reflection and photoluminescence spectra of ultrathin MoS_2 layers. (a) Reflection difference due to an ultrathin MoS_2 layer on a quartz substrate, which is proportional to the MoS_2 absorption constant. The observed absorption peaks at 1.85 eV (670 nm) and 1.98 eV (627 nm) correspond to the A and B direct excitonic transitions with the energy split from valence band spin-orbital coupling. The inset shows the bulk MoS_2 band structure neglecting the relatively weak spin-orbital coupling, which has an indirect bandgap around 1 eV and a single higher energy direct excitonic transition at the *K* point denoted by an arrow. (b) A strong photoluminescence is observed at the direct excitonic transitions energies in a monolayer, respectively [30]. (Copyright 2010 by American Chemical Society, reprinted with permission).

2.4. Defect induced photoluminescence and single photon source

As known, vacancy defects, impurities, potential wells created by structural defects or local strain or other disorders might be introduced in the growth process of the TMDC materials [12, 13, 32, 33]. They can produce localized states to participate the optical emission and absorption as manifested by P_1 to P_3 in **Figure 10**, and the emission bands on the lower energy side of the peak XX in **Figure 11**. Since the point defects can induce intervalley coupling, the defect-related emission peaks show no measurable circular polarization character. Besides, the excitons, trions, and even biexcitons can be trapped by theses crystal structure imperfections to form corresponding bound quasiparticles. Therefore, delocalized excition, charged exciton and biexciton emissions, and localized ones can coexist in the TMDCs. Interestingly, these carrier trapping centers can act as single-photon emitters to emit stable and sharp emission line [12, 13, 32, 33]. For this kind of single quantum emitter, since the maximum number of

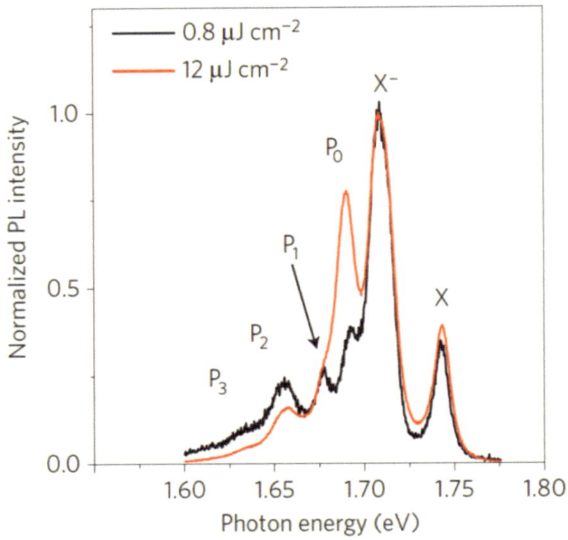

Figure 10. Photoluminescence spectra (PL) of monolayer WSe_2 at 50 K for pulsed excitation under applied pump fluences of 0.8 μJ cm^{-2} (black curve) and 12 μJ cm^{-2} (red curve), respectively. The spectra are normalized to yield the same emission strength for the neutral exciton [40]. (Copyright 2015 by the Nature Publishing Group. Reprinted with permission.)

emitted single photons is limited by the lifetime of the excited state, a saturation of the PL intensity at high excitation laser power is expected.

2.5. Magneto-optical properties of the monolayer TMDCs

The presence of a magnetic field induces a quantization of the energy levels. At high magnetic field, the Landau levels (LLs) form. The transition rate between the conduction–and valence–band Landau levels can be calculated using the eigenfunctions in Eq. 21.

$$\langle \Psi_n^{\tau,+} | \mathrm{H}_{L-M}^{CP,\pm} | \Psi_m^{\tau,-} \rangle = M\, \delta_{szc,\,szv}\, [(\tau+\sigma)\, c_{1,n}^{\tau,+}\, c_{2,m}^{\tau,-}\, \delta_{n,\,m+\tau} + (\tau-\sigma)\, c_{2,n}^{\tau,+}\, c_{1,m}^{\tau,-}\, \delta_{n+\tau,\,m}] \tag{30}$$

where

$$M = \frac{etaA_0}{2\hbar N_\tau^{n,+} N_\tau^{m,-}}. \tag{31}$$

In the presence of magnetic field, both σ^+ and σ^- absorptions take place in each valley. However, there is a great difference in their intensity. For instance, the absorption spectrum intensity of σ^+-light is 10^4 times larger than σ^--light in the *K*–valley. And, the lowest transition energy absorption spectrum demonstrates a valley polarization, i.e., σ^+ in the *K*–valley and σ^- in the K'–valley, as showed in **Figure 12** [41]. In addition, the transition occurs from $n = 0$ to $n = 1$ LLs in the K'–valley, whereas $n = 1$ to $n = 0$ LLs in the *K*–valley. Therefore, the valley

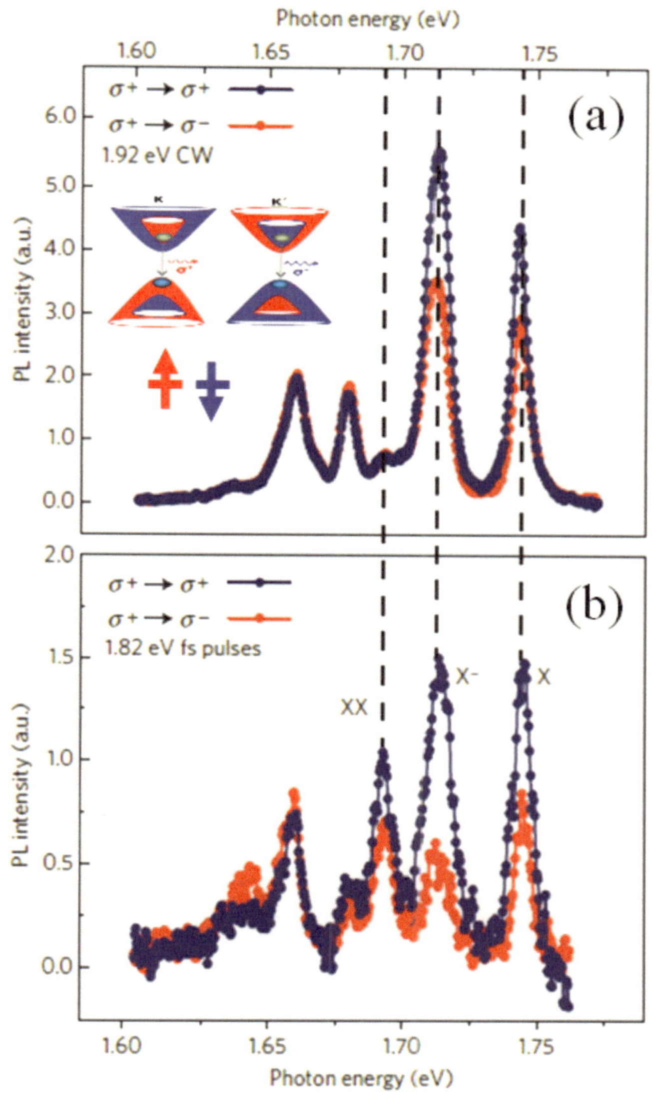

Figure 11. Circularly polarized photoluminescence (PL) spectra of monolayer WSe_2 excited by near-resonant circularly polarized radiation at 15 K. (a) PL for low exciton density with continuous wave excitation at a photon energy of 1.92 eV. (b) PL for high exciton density with pulsed excitation at a photon energy of 1.82 eV. Blue and red curves correspond to the same and opposite circularly polarized states. The emission energies for neutral (X) and charged (X^-) excitons and the biexciton (XX) state are indicated by dashed lines. Inset shows schematic diagram of energy dispersion in K- and K'-valleys and valley-polarized emission. The vertical arrows indicate the electron spin direction. The circles represent conduction and valence band electrons [40]. (Copyright 2015 by the Nature Publishing Group. Reprinted with permission).

polarization remains in the magneto-optical absorption, as showed in **Figure 13**. It is worth to argue that the higher-order terms in the effective **k.p** model only induces about 0.1% correction to absorption spectrum intensity [41], which allows us neglect them safely.

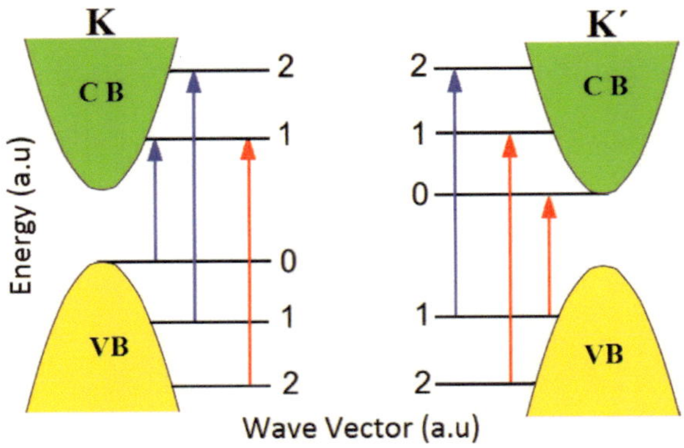

Figure 12. Selection rules for the interband transitions between Landau levels in K-valley (the left panel) and in K'-valley–valley (the right panel) of monolayer WS_2 subjected to a magnetic field along the $\hat{z}$-direction, excited by a circularly polarized light. The blue and red arrows correspond to σ^+ and σ^- absorptions, respectively [41]. (Copyright 2013 by the American Physical Society. Reprinted with permission).

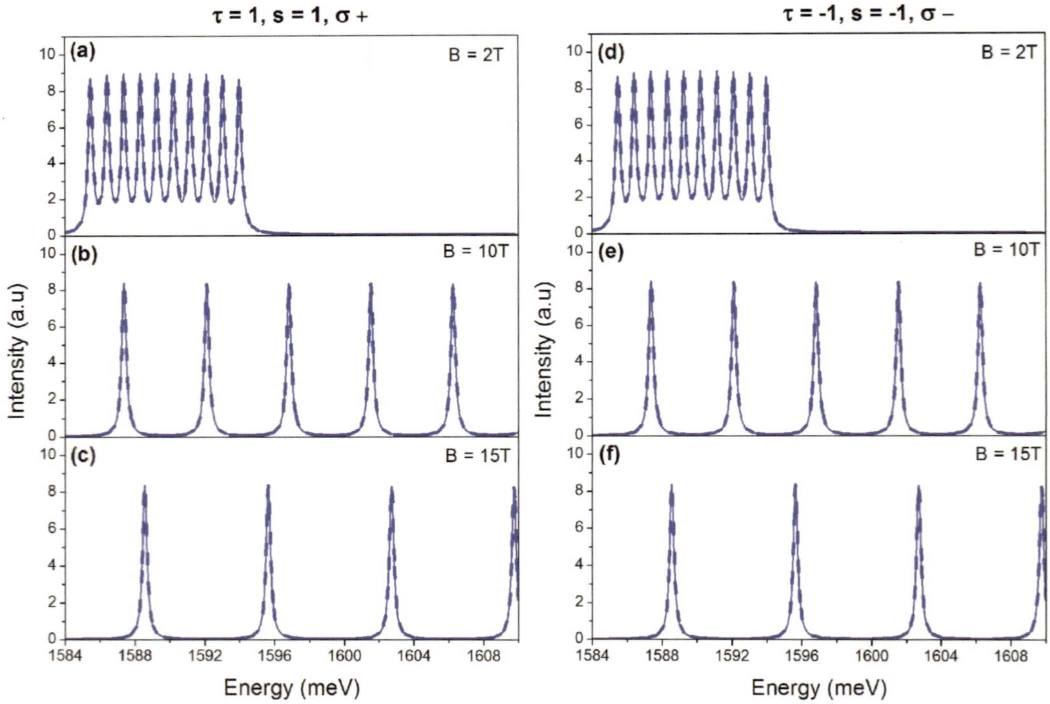

Figure 13. σ^+ absorption spectrum of monolayer MoS_2 for spin–up states in the K-valley (the left panels) and σ^- absorption spectrum for spin–down states in the K'-valley (the right panels), for a magnetic field $B = 2$ ((a) and (d)), 10 ((b) and (e)), $15T$ ((c) and (f)), respectively.

2.6. TMDC quantum dots and valley polarized single-photon source

The Hamiltonian of the TMDC QDs in polar coordinates is given by [15]

$$\mathrm{H} = \begin{pmatrix} \frac{\Delta}{2} & tae^{-i\tau\theta}\left(-i\tau\frac{\partial}{\partial_r} - \frac{1}{r}\frac{\partial}{\partial_\theta}\right) \\ tae^{i\tau\theta}\left(-i\tau\frac{\partial}{\partial_r} + \frac{1}{r}\frac{\partial}{\partial_\theta}\right) & -\frac{\Delta}{2} + \tau s\lambda \end{pmatrix}. \tag{32}$$

As a matter of convenience, we get rid of the angular part by using the following *ansatz* for the eigenfunctions

$$\boldsymbol{\Psi} = \begin{pmatrix} \psi_a(r,\theta) \\ \psi_b(r,\theta) \end{pmatrix} = \begin{pmatrix} e^{im\theta}\overline{a}(r) \\ e^{i(m+\tau)\theta}\overline{b}(r) \end{pmatrix}, \tag{33}$$

where the quantum number $m = j - \tau/2$, j is the quantum number related to the effective angular momenta $J_{eff}^{\tau} = L_z + \hbar\tau\sigma_z/2$ with L_z being the orbital angular momenta in the $\hat{z}$ direction. Then the Schrödinger equation for TMDC QDs, showed in **Figure 14**, becomes

$$\begin{aligned} \left(\frac{\Delta}{2} - E\right)\overline{a}(r) &= ita\left(\tau\overline{b}'(r) + \overline{b}(r)\frac{(m+\tau)}{r}\right), \\ \left(\frac{\Delta}{2} - \tau s\lambda + E\right)\overline{b}(r) &= ita\left(-\tau\overline{a}'(r) + \overline{a}(r)\frac{m}{r}\right). \end{aligned} \tag{34}$$

After some algebra calculations, we get two decoupled equations. They are

$$\overline{a}''(r) + \frac{\overline{a}'(r)}{r} + \overline{a}(r)\left(\chi - \frac{m^2}{r^2}\right) = 0 \tag{35}$$

and

$$\overline{b}''(r) + \frac{\overline{b}'(r)}{r} + \overline{b}(r)\left(\chi - \frac{(m+\tau)^2}{r^2}\right) = 0 \tag{36}$$

where $\chi = \frac{(2E-\Delta)(\Delta+2E-2\lambda s\tau)}{4t^2a^2}$. These two second order differential equations can be straightforwardly resolved. Finally we obtain,

$$\overline{a}(r) = \overline{N}\left(\frac{2iat\sqrt{\chi}}{\Delta - 2E}\right)J_{|m|}(r\sqrt{\chi}) \tag{37}$$

and

$$\overline{b}(r) = \overline{N}J_{|m+\tau|}(r\sqrt{\chi}), \tag{38}$$

where $\overline{N}$ is the normalization constant and $J_n(x)$ is Bessel Function of the first kind. Applying the infinite mass boundary condition $\frac{\psi_2(R,\theta)}{\psi_1(R,\theta)} = i\tau e^{i\tau\theta}$, we obtains the secular equation

$$J_{|m+\tau|}(R\sqrt{\chi}) = -\frac{\tau(2at\sqrt{\chi})J_{|m|}(R\sqrt{\chi})}{\Delta - 2E}, \tag{39}$$

where R is the QD radius, see **Figure 14**.

From **Figures 15** and **16**, we see that the bound states formed in a single valley, and $E^{\tau}(j) \neq E^{\tau}(-j)$ for both conduction– and valence– bands. In addition, the electron-hole symmetry is broken. Due to the confinement potential, the effective time reversal symmetry (TRS) is broken within a single valley, even without the magnetic field, similarly to graphene QDs [42]. On the other hand, the inverse asymmetry of the crystalline structure, the terms of spin-orbit interaction and the confinement potential, do not commute with the effective inversion operator, defined in a single valley $P_e = I_\tau \otimes \sigma_x$. Consequently, the QDs do not preserve the electron-hole symmetry in the same valley. However, comparing **Figure 15(a, b)**, we can found that $E^{\tau}(j) = E^{-\tau}(-j)$ was still true. This is attributed to the TRS, where $THT^{-1} = H$, the TRS operator is defined as $T = i\tau_x \otimes s_y C$, with C being the conjugate complex operator.

2.7. Landau levels in monolayer MoS_2 quantum dots

Similarly to what we did in the case of the monolayer TMDCs, for quantum dot subjected to a perpendicular magnetic field, we do the Peierls substitution and use now the symmetric

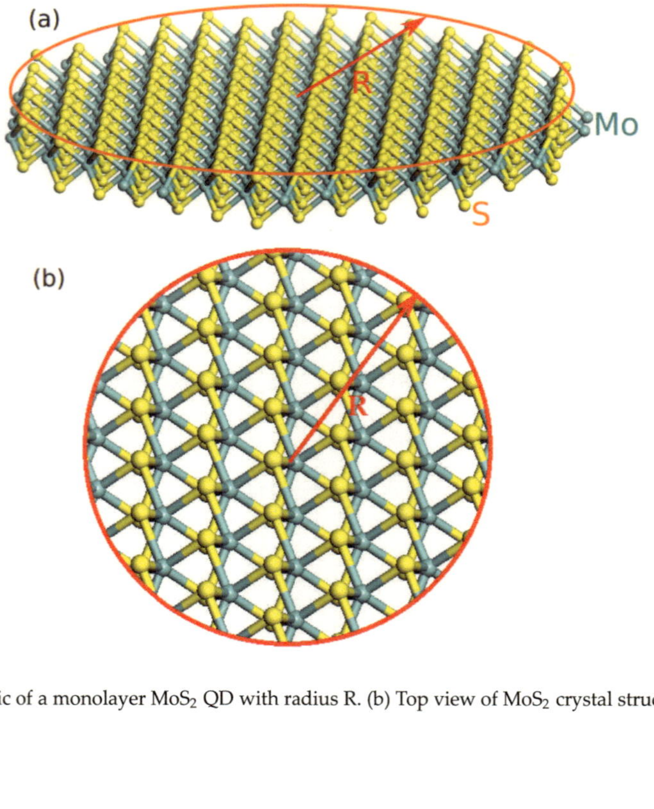

Figure 14. (a) Schematic of a monolayer MoS_2 QD with radius R. (b) Top view of MoS_2 crystal structure.

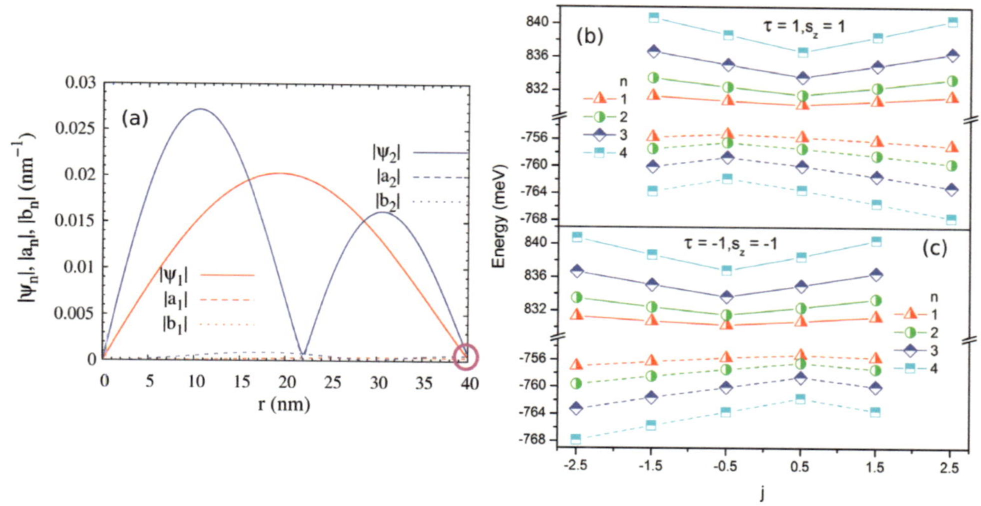

Figure 15. (a) Wave function profile Ψ_n and its two components (a_n, b_n) of the lowest two (n = 1, 2) conduction band states in the K–valley with spin–up at j = 1.5. The pink circle indicates that the wave function Ψ_n is nonzero (even though small) at r = R. Energy spectrum of the lowest four (n = 1, 2, 3, 4) conduction bands (solid curves) and the highest four valence bands (dashed curves) as a function of j, in the K-valley with spin-up (b) and K'-valley with spindown (c) of a MoS_2 dot. The dot size is R = 40nm and the magnetic field is B = 0 [14]. (Copyright 2016 by the IOP publishing. Reprinted with permission).

gauge, i.e., $\vec{A}= (-By/2,Bx/2)$. The part of magnetic field Hamiltonian, in polar coordinates is given by

$$\mathcal{H}_B = \begin{pmatrix} 0 & \frac{ta}{2l_B^2}\left(-ie^{-i\tau\theta}r\right) \\ \frac{ta}{2l_B^2}\left(ie^{i\tau\theta}r\right) & 0 \end{pmatrix}. \tag{40}$$

Then, the total Hamiltonian becomes $H + \mathcal{H}_{B,}$ which leads to the following two-coupled differential equations:

$$\left(\frac{\Delta}{2} - E\right)\bar{a}(r) = iat\left(\tau\bar{b}'(r) + \bar{b}(r)\left(\frac{(m+\tau)}{r} + \frac{r}{2l_B^2}\right)\right),$$

$$\left(E + \frac{\Delta}{2} - \lambda s\tau\right)\bar{b}(r) = iat\left(-\tau\bar{a}'(r) + \bar{a}(r)\left(\frac{m}{r} + \frac{r}{2l_B^2}\right)\right). \tag{41}$$

In order to solve this eigenvalue problem, let us first decouple these two equations into

$$\bar{a}''(r) + \frac{\bar{a}'(r)}{r} + \bar{a}(r)\left(-\frac{r^2}{4l_B^4} - \kappa - \frac{m^2}{r^2}\right) = 0 \tag{42}$$

and

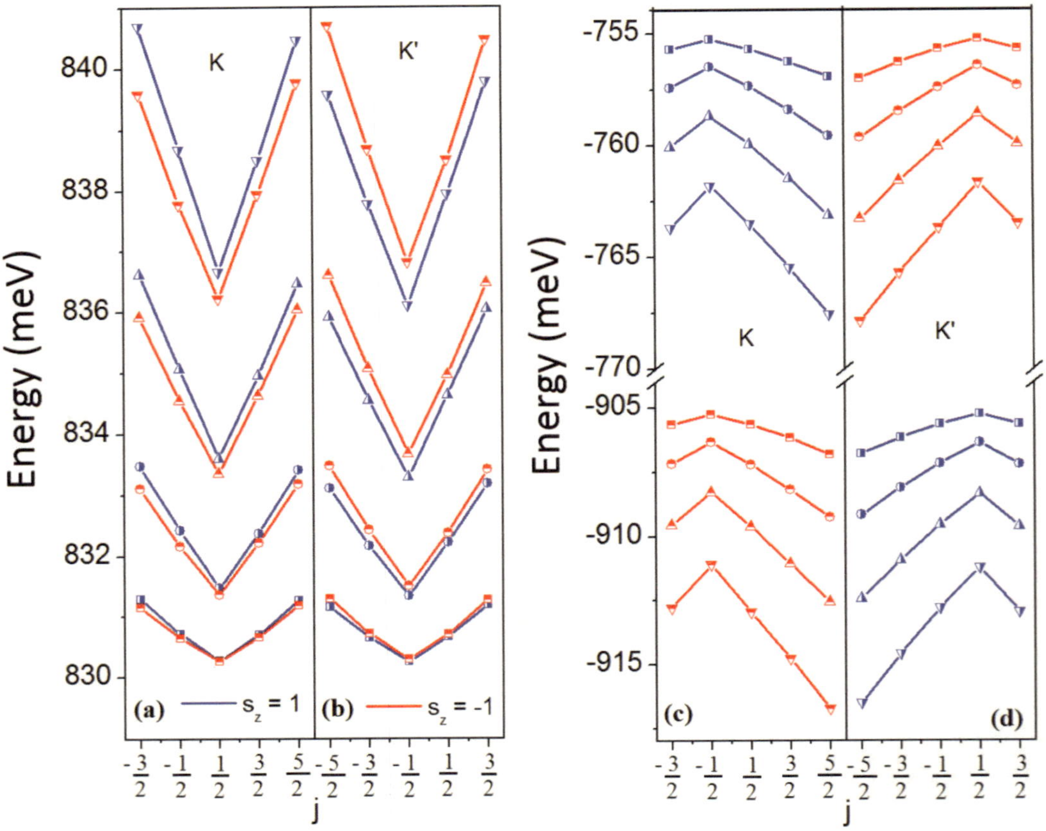

Figure 16. Energy spectrum of monolayer MoS_2 QD with a radius of 40 nm for the conduction–band in the K- (a) and K'- (b) valleys. (c) and (d) are corresponding figures for the valence–band. The blue and red curves correspond to spin–up and spin–down energy levels, respectively.

$$\bar{b}''(r)+\frac{\bar{b}'(r)}{r}+\bar{b}(r)\left(-\frac{r^2}{4l_B^4}-\phi-\frac{(m+\tau)^2}{r^2}\right)=0 \tag{43}$$

where

$$\kappa=\frac{(m+\tau)}{l_B^2}+\frac{(\Delta-2E)(\Delta+2E-2\lambda s\tau)}{4t^2a^2}, \tag{44}$$

$$\phi=\frac{m}{l_B^2}+\frac{(\Delta-2E)(\Delta+2E-2\lambda s\tau)}{4t^2a^2} \tag{45}$$

Solving these equations, we obtain the following two components of the eigenfunctions

$$\bar{a}(r)=G_\xi\overline{N}r^{|m|}e^{-\frac{r^2}{4l_B^2}}{}_1F_1\left(\frac{1}{2}\left(\kappa l_B^2+|m|+1\right);|m|+1;\frac{r^2}{2l_B^2}\right), \tag{46}$$

$$\overline{b}(r) = \overline{N} r^{|m+\tau|} e^{-\frac{r^2}{4l_B^2}} {}_1F_1\left(\frac{1}{2}\left(\phi l_B^2 + |m+\tau| + 1\right); |m+\tau| + 1; \frac{r^2}{2l_B^2}\right) \tag{47}$$

where

$$G_\xi = iW(\tau m)\left(\frac{4ta\left(m + \tau Q_+(\tau m)\right)}{\Delta - 2EW(\tau m) - 2\tau s\lambda Q_-(\tau m)}\right)^{W(\tau m)}, \tag{48}$$

$$W(\tau m) = S(\tau m + \varepsilon), \tag{49}$$

$$Q_\pm(\tau m) = \frac{1 \pm S(\tau m + \varepsilon)}{2}, \tag{50}$$

$S(x)$ is the sign function, ${}_1F_1(a, b, x)$ is the confluent hypergeometric function of the first kind, ε is an arbitrary constant ($0 < \varepsilon < 1$) used to avoid a singularity in the function $S(x)$.

With the eigenfunctions at hand, we can derive the secular equation for the eigenvalues by applying infinite mass boundary condition, i.e.,

$$\frac{{}_1F_1\left(\frac{1}{2}\left(\phi l_B^2 + |m+\tau| + 1\right); |m+\tau| + 1; \frac{R^2}{2l_B^2}\right)}{{}_1F_1\left(\frac{1}{2}\left(\kappa l_B^2 + |m| + 1\right); |m| + 1; \frac{R^2}{2l_B^2}\right)} = i\tau R^{|m| - |m+\tau|} G_\xi \tag{51}$$

Figure 17(a) illustrates the energy spectrum of the lowest four spinup conduction bands in the K-valley ($\tau = 1$, $s = 1$) as a function of magnetic field (B), for the orbital angular momentum $m = \pm 0, 1, \pm 2$, in the 70 nm dot. At zero magnetic field, the electronic shells of an artificial atom such as s, p, and d shells emerge. In a certain valley, say the K-valley, the atomic states possess both spin- and orbital–degeneracies such as $E(m) = E(-m)$. In addition, we should emphasize that the energy spectrum in different valleys have time reversal symmetry at $B = 0$. However, the presence of magnetic field breaks down this symmetry and leads to splittings of the atomic orbitals of the dot. Moreover, in the regime of weak magnetic fields, unlike the monolayer MoS_2 in which the energy shows linear B response, valley dependent energy levels with nonlinear B response are observed. Such effect is attributed to the competition of the QD confinement with the magnetic field effect [14].

As B increases, an effective confinement induced by the magnetic field gradually becomes comparable to that of the dot. Hence, their contributions to the electronic energy are balanced. With a further increasing of B, magnetic field effect starts to dominate the features of the energy spectrum. Accordingly, the LLs which show a linear dependence on B, became of the heavily massive Dirac character, are formed just like in the pristine monolayer MoS_2. The higher the energy level is, the stronger the magnetic field needed to form the corresponding LL. For instance, the lowest LL is formed around a critical value $B = B_c = 2T$. Interestingly, in the Hall regime, an energy locked (energy independent on B) mode referring to the lowest

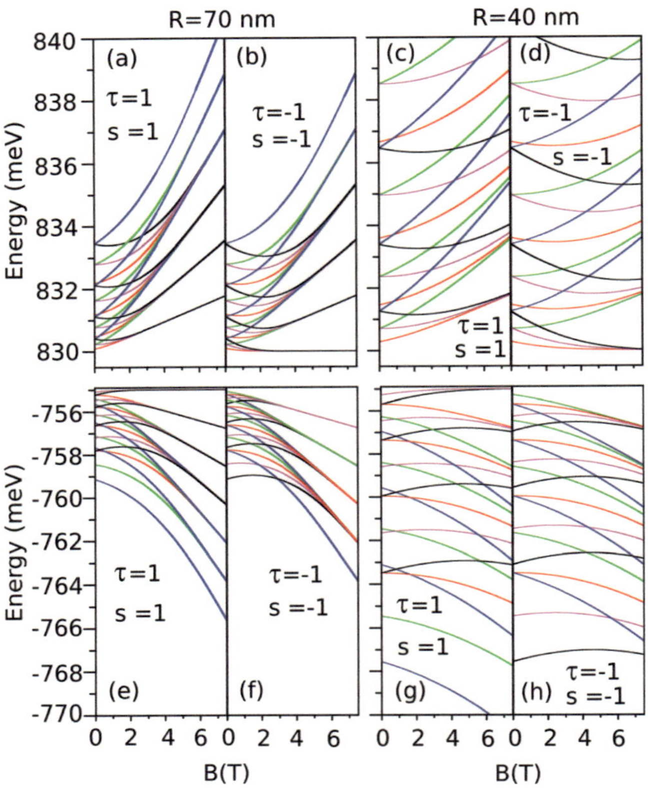

Figure 17. Energy spectrum of the lowest four conduction band states with spin–up in the *K*-valley (a) and spindown in the *K*′-valley (b), and the highest four valence band states with spin–up in the *K*-valley (e) and spin-down in the *K*′-valley (f) of a 70 nm MoS_2 QD, as a function of the magnetic field, for angular momentum $m = 0$ (red curves), -1 pink curves), 1 (green curves), -2 (black curves), and 2 (blue curves). The corresponding analogues for a 40 nm dot are shown in (c), (d), (g), and (h), respectively [14]. (Copyright 2016 by the IOP publishing. Reprinted with permission).

spin-down conduction band in the *K*′-valley emerges, as shown in **Figure 17(b)**. It is expected to be an analog to the zero energy mode in gapless graphene, associating with certain topological properties. These novel features of the QD energy spectrum are tunable by QD size [14].

Figure 17(c, d) are the corresponding analogs of **Figure 17(a, b)**, but for a dot with R = 40 nm. Comparing with the 70nm dot ($B_c = 2T$), here the energy locked mode takes place until $B_c = 4T$, which turns out to be larger than the one (2 T) for the 70 nm dot, indicating the locked energy modes arise from the competition between the dot confinement and the applied magnetic field [14].

Let us turn to the energy spectrum of the valence band in the dot of $R = 70nm$. **Figure 17(e, f)** show the B field dependence of the energy spectrum of the highest four valence bands for several values of angular momentum ms with $m = 0, \pm 1, \pm 2$ in the *K*-valley with spin-up ($\tau = 1, s = 1$) and *K*′-valley with spindown ($\tau = -1,\ s = -1$), respectively. In contrast to the conduction band, here we find $E(m) \neq E(-m)$ even at B = 0 due to the spin–orbit coupling, see **Figure 17(a, b, e, f)**. Besides that, we also observe the emergent locked energy mode around

$B_c = 2T$, but referring to the highest valence band. We should emphasize that the locked energy modes for the conduction and valence bands appear in distinct valleys with opposite spin, see **Figure 17(b, e)**. The corresponding valence band energy spectra for the 40 nm dot are shown in **Figure 17(g–h)**, where the locked energy mode associated with a larger $B_c = 4T$, also arises from the combined effect of the dot confinement and the applied magnetic field, similar to the conduction band. Therefore, the flat band or energy locked modes appear only in the valence band of the K-valley and the conduction band of the K'-valley [14]. The effect of changing the magnetic field direction is showed in **Figure 19**.

A comparison of the energy spectrum of the 70 nm dot with that of the bulk TMDC (i.e., infinite geometry) is shown in **Figure 18**. Because of the large effective mass at the band edges, the LLs of the bulk TMDC scale as $E_{\pm}(\omega_c, n) = \frac{\lambda \tau s}{2} \pm \sqrt{\frac{(\Delta - \lambda \tau s)^2}{4} + t^2 a^2 \omega_c^2 n}$ in the low energy region, which resembles conventional 2D semiconductors more than Dirac fermions. The $n = 0$LL appears only in the conduction band of K'-valley (and the valence band of K-valley, not shown), implying the lifting of valley degeneracy for the ground state. As magnetic field increases, there is an evolution of the energy spectrum from atomic energies to Landau levels. More specifically, in the regime of weak B field, the atomic structure emerges, where the energy is distinct from that of the bulk case. On the other hand, in the strong field regime, the energy in the bulk TMDC and quantum dot becomes identical because the effective confinement due to the magnetic field dominates physical behaviors of the dot. Note that although the two band model we used here is widely adopted in the literature [39], the model itself still has limitations, e.g., it cannot properly describe the spin splitting of the conduction band, the trigonal warping of the spectrum and the degeneracy breakdown by applied magnetic field

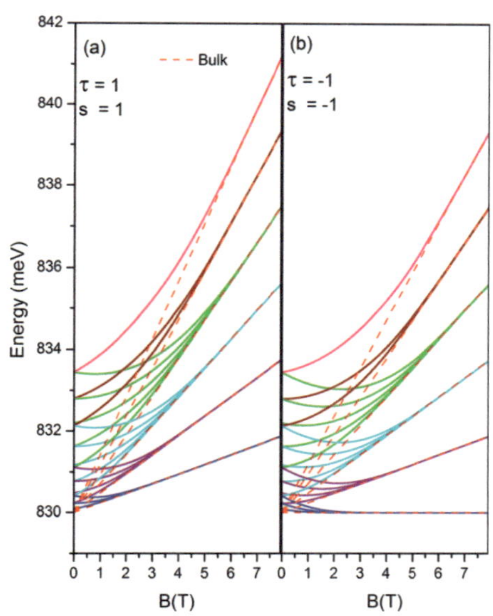

Figure 18. Energy spectrum of the conduction band states with spin–up in the K-valley (a) and spin-down in the K'-valley (b) of a monolayer MoS_2 (dotted curves) and of a MoS_2QD with radius $R = 70$ nm, as a function of the magnetic field.

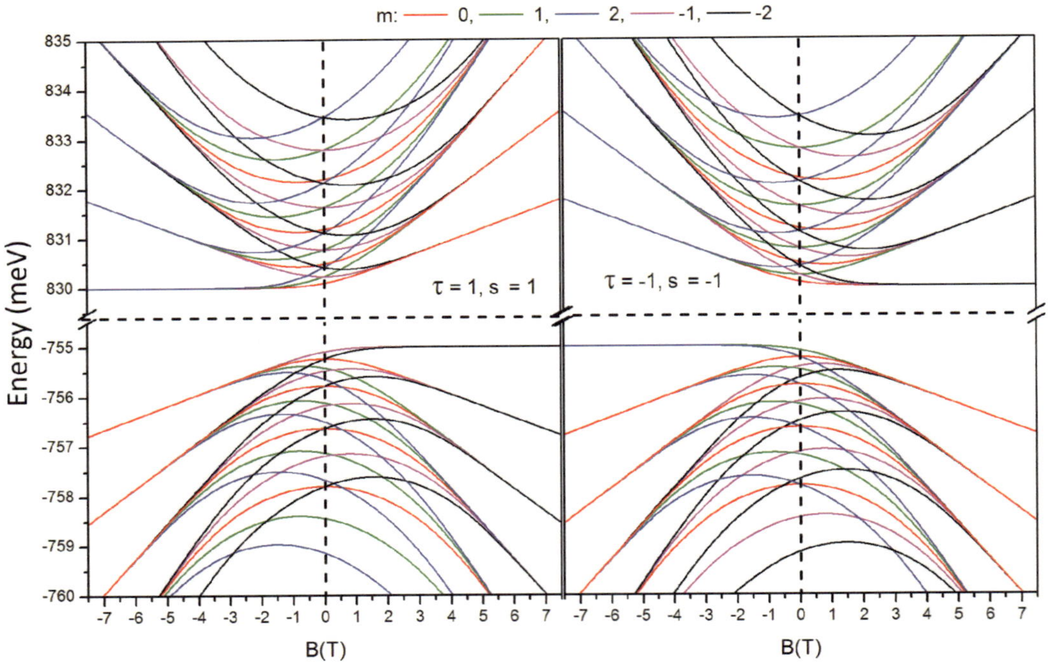

Figure 19. Energy spectrum of the states with spin-up in the K-valley (a) and spindown in the K'-valley (b) of a monolayer MoS_2 QD with radius $R = 70$ nm, as a function of the magnetic field along both positive and negative $\hat{z}$–direction, for angular momentum $m = 0$ (red curves), -1 (pink curves), 1 (green curves), -2 (black curves), and 2 (blue curves).

for Landau levels with the same quantum number. However, for usual experimental setups, these effects in the vicinity of the K or K' valley in which we are interested play a minor role. Hence it can be safely neglected.

2.8. Optical selection rules in monolayer MoS_2 quantum dots

In the QDs as demonstrated in **Figure 20**, the optical transition matrix elements in the QDs are computed by,

$$\langle \Psi_c | \mathcal{H}_{L-M} | \Psi_v \rangle = \left(\frac{\pi A_0}{\hbar} \right) \delta_{s_{zv}, s_{zc}} [(\tau - \sigma) \delta_{m_v, m_c + \tau} R_{-\sigma} + (\tau + \sigma) \delta_{m_c, m_v + \tau} R_\sigma)], \tag{52}$$

where s_{zv} (s_{zc}) denote the valence (conduction) band spin state, $R_{-\sigma} = \int_0^R b_c^* a_v r dr$ and $R_\sigma = \int_0^R a_c^* b_v r dr$ with $a_{c/v}$ and $b_{c/v}$ the radial components of the conduction/valence band spinor. The selection rule for optical transitions in TMDC QDs is defined by $m_v - m_c = \pm\tau$ and $s_{zv} = s_{zc}$, i.e., the angular momentum of the initial and final states differs by ± 1, but the spin of these two states is the same. The magnitude of transition rates is determined by the integral $R_{-\sigma}$ and R_σ for the transitions taking place in the valley $\tau = -\sigma$ and σ, respectively. Since $|a_c(r)| > |b_c(r)|$ and $|a_v(r)| < |b_v(r)|$, the integral $R_{-\sigma}$ is much smaller than the R_σ. As a

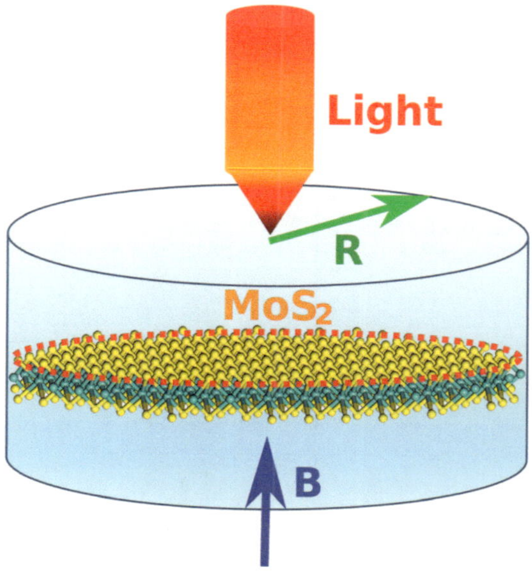

Figure 20. Schematic of a monolayer MoS_2 circular quantum dot with radius R indicated by a red circle, excited by a light field. A magnetic field B is applied perpendicularly to the MoS_2 sheet [15]. (Copyright 2017 by the Nature Publishing Group. Reprinted with permission).

consequence, the absorption in the valley $\tau = \sigma$ is stronger than that in the $\tau = -\sigma$, which leads to a valley selected absorption. In a special case in which one component of the wave function spinor is equal to zero such as $n_{II} = 0$ LL, only photons with the helicity $\sigma = \tau$ is absorbed. Then, one obtains a dichroism $\eta = 1$. For the linear polarized light, however, there is no valley polarization in the absorption spectrum, in contrast to the case of CPL [15].

In the 2D bulk MoS_2, the bottom of the conduction band at the two valleys is characterized by the orbital angular momentum $m = 0$. In contrast, at the top of the valence band, the orbitals with $m = 2$ in the K-valley, whereas m is equal to -2 in the K'-valley. The valley dependent angular momentum in the valence band allows one to address different valleys by controlling the photon angular momentum, i.e., the helicity of the CP light. Such valley-specific circular dichroism of interband transitions in the 2D bulk has been confirmed [19, 43, 44]. The question is whether its counterpart QD also possesses this exotic optical property. **Figure 21(a, b)** depict zero-field band-edge optical absorption spectrum of a 70-nm dot pumped by the CP light field. Interestingly, one notices that (i) the polarization of the absorption spectrum is locked with the valley degree of freedom, manifested by the intensity of absorption spectrum with $\sigma = \tau$ being about 106 times stronger than that with $\sigma = \tau$, and (ii) the spectrum is spin-polarized. Thus, the QDs indeed inherit the valley and spin dependent optical selection rule from their counterpart of 2D bulk MoS_2. In spite of the distinction in the spin- and valley-polarization of absorption spectra in the distinct valleys, their patterns are the same required by the time reversal symmetry. In **Figure 21(c, f)** we show the zero-field optical absorption spectrum as a function of excitation energy for several values of dot-radius within R = 20–80 nm. The involved transitions in **Figure 21(c, f)** lagged by the numbers have been schematically illustrated in **Figure 21(g)**. Alike conventional semiconductor QDs, several peaks stemmed from discrete excitations of

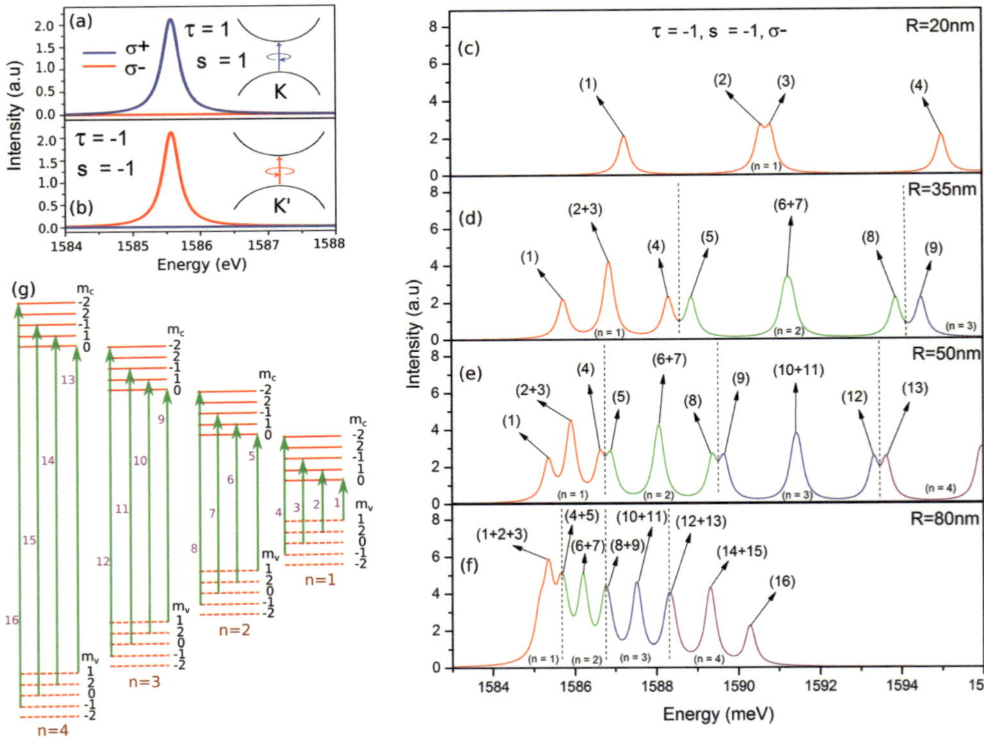

Figure 21. Zero-field optical absorption spectrum associated with interband transitions between conduction and valence-band ground states of a 70-nm MoS_2 dot for the spin-up state in the *K*-valley (a) and for the spin-down state in the *K'*-valley (b), pumped by both clockwise (σ^+, blue curve) and anticlockwise (σ^-, red curve) circularly polarized light fields. (c–f) Absorption intensity for the spin-down state in the *K'*-valley under the excitation of σ^-, for QDs of R = 20, 35, 50, and 80 nm, respectively. (g) Schematic diagram of the involved interband transitions in (c-f) tagged by the numbers. The color of the curves is used to highlight the principle quantum number (n) of transition involved states [15]. (Copyright 2015 by the Nature Publishing Group. Reprinted with permission).

the MoS_2 QD are observed. As the dot size decreases, the peaks of absorption spectrum undergo a blue shift. In other words, a reduction of the dot size pushes the electron excitations to take place between higher energy states, as a result of the enhancement of the confinement on carriers induced by a shrink of the dot. Therefore, the spin-coupled valley selective absorption with a tunable transition frequency can be achieved in QDs by varying dot geometry, in contrast to the 2D bulk where a fixed transition frequency is uniquely determined by the bulk band structure. In addition to the blue shift of the transition frequency, the absorption intensity can also be controlled by dot geometry due to size dependence of the peak interval. In fact, an increasing of dot size results in a reduction of the energy separation among confined states. Thus, as the dot size increases, the absorption peaks get closer and closer, see **Figure 21(c–f)**. Eventually, several individual absorption peaks merge together to yield a single composite-peak with an enhanced intensity. For instance, for the 20-nm dot (**Figure 21c**), the lowest energy peak is generated by only one transition, labeled by (1) (see also **Figure 21g**). However, for the dot with R = 80nm, we observe a highly enhanced absorption intensity labeled by

$(1+2+3)$, which in the 20-nm dot refers to three separated peaks with weak absorption intensities indicated by (1), (2), (3), respectively.

2.9. Magneto-optical properties of monolayer MoS_2 quantum dots

With the knowledge of the energy spectrum and eigenfunctions of the QD, we are ready to study its magneto-optical properties. The optical transition matrix in Eq. 27 is applicable to the current case provided that we use newly obtained wavefunctions presented in Eq. 47 and Eq. 48.

Figure 22 shows the magneto-optical absorption spectra for the spin-down states in the K'-valley of a monolayer MoS_2 QD with $R = 40$ nm excited by left-hand circularly polarized

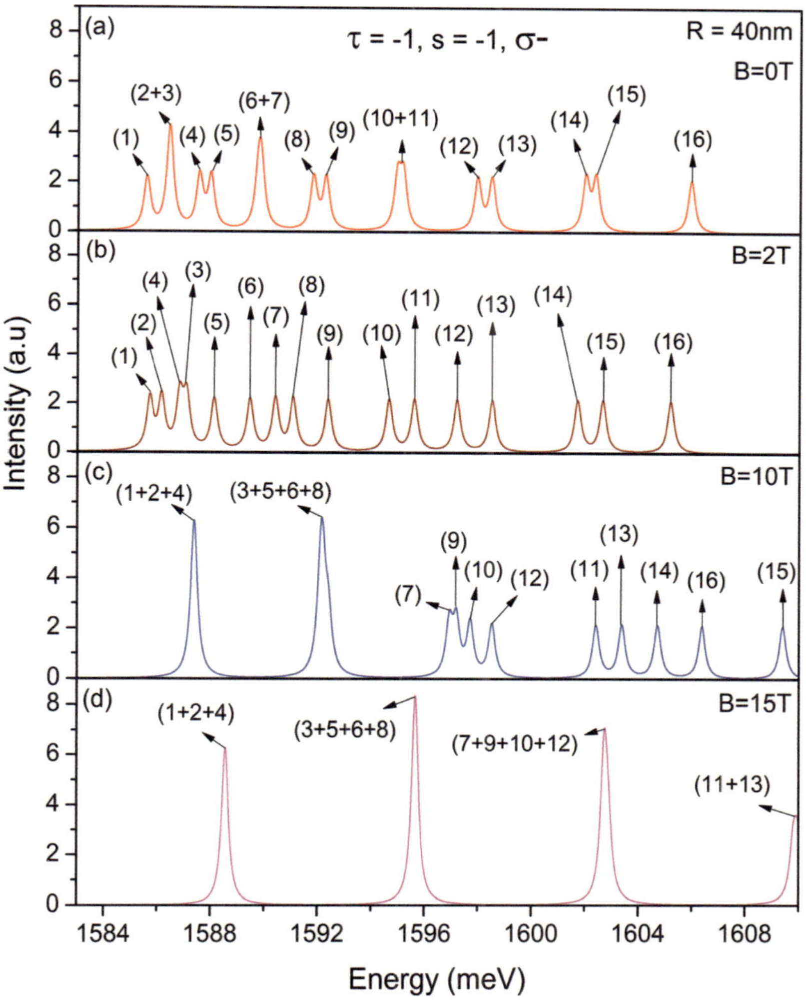

Figure 22. Absorption spectrum for the spin-down states in the K'-valley of a monolayer MoS_2 QD with $R = 40$ nm excited by left-hand circularly polarized light σ^-, at the magnetic field $B = 0$ (a), 2T (b), 10T (c), 15T (d), respectively. The corresponding enumerated optical transitions are schematically shown in **Figure 21(g)** [15]. (Copyright 2015 by the Nature Publishing Group. Reprinted with permission).

light σ^-, for several values of magnetic field ranging from 0 to 15 T. A few interesting features are observed. Firstly, the magneto-optical absorption is also spin- and valley-dependent, as it does in optical absorption spectrum at zero field. In particular the lowest transition energy absorption peak related to the interband transition involving $n_{ll} = 0$ LL is totally valley polarized with dichroism equal to 1. Thus, the polarization of magneto-optical absorption locks with the valley. Secondly, for a fixed value of dot size, increasing (decreasing) the strength of the magnetic field results in a blue (red) shift in the absorption spectrum. This arises from the fact that the magnetic field induces an effective confinement characterized by the magnetic length $l_B = \sqrt{\hbar/eB}$, which is more pronounced for a stronger magnetic field. In addition, the magnetic quantization induced by the magnetic field favors the QD to absorb photons with higher energies. The stronger the magnetic field, the greater the capacity of the QDs to absorb the photons of higher energy. Thirdly, in the high magnetic-field regime the absorption intensity can be highly enhanced by increasing the strength of magnetic field due to increased degeneracy of the LLs.

2.10. Excitonic effect in monolayer MoS_2 quantum dots

The optical and magneto-optical absorptions that we have discussed so far are based on the independent electron-hole picture. In reality, there is a strong Coulomb interaction between the electron and hole in MoS_2. The full treatment of the Coulomb interaction using many-body theory is beyond the scope of this chapter. However, the excitonic effects in absorption spectrum of the monolayer MoS_2 QDs can be properly addressed by an exact diagonalization, which is adopted in this work. The Hamiltonian used to describe the exciton is, $H(\mathbf{r}_e, \mathbf{r}_h) = \mathcal{H}_e(\mathbf{r}_e) + \mathcal{H}_h(\mathbf{r}_h) + V^{e-h}(\mathbf{r}_e - \mathbf{r}_h)$, where $\mathcal{H}_{e(h)}$ is the single electron (hole) Hamiltonian in QDs (see Eq. 1) and V^{e-h} is the electron-hole Coulomb interaction given by $V^{e-h}(|\mathbf{r}_e - \mathbf{r}_h|) = (1/4\pi\epsilon_r\epsilon_0)(e^2/|\mathbf{r}_e - \mathbf{r}_h|)$. Here ϵ_0 is the permittivity, ϵ_r is the dielectric constant, and $\mathbf{r}_e$ and $\mathbf{r}_h$ respectively stand for the position of electron and hole. An exciton can be understood as a coherent combination of electron-hole pairs. Thus, the wavefunctions of an exciton can be constructed based on a direct product of single-particle wave functions for the electron and hole. Since the electron-hole Coulomb interaction is larger than or at least in the same magnitude as the quantum confinement in the MoS_2 QDs, the excitonic effect play an very important role. Hence the single-particle wavefunctions should be modified due to the presence of the Coulomb interaction. Therefore, to get quick convergence of numerical calculation, instead of using simple single-particle wavefunction, we use the modified one to construct the exciton states, i.e., $\chi_j(r_{e,h}) = N\exp\left(-r_{e,h}/r_b\right)\mathbf{\Psi}_j(r_{e,h})$, where the exponential factor is a hydrogen-like s-wave state, N is the normalization constant, $\mathbf{\Psi}_j$ is the wave function of the Hamiltonian $\mathcal{H}_{e,h}$, and r_b is the exciton bohr radius, which has the value of $r_b \sim 1$ nm in MoS_2 [24]. Then, the exciton wave function $\mathbf{\Psi}_{exc}$ can be straightforwardly written as,

$$\mathbf{\Psi}^{v}_{exc}(\mathbf{r}_e, \mathbf{r}_h) = \sum_{i,j} C^{v}_{i,j}\ \chi_i(\mathbf{r}_e)\,\chi_j(\mathbf{r}_h), \tag{54}$$

with the superscript ν referring to the ν-th exciton state. Within the Hilbert space made of states of electron-hole pairs $\{\chi_i(\mathbf{r}_e)\chi_j(\mathbf{r}_h)\}$, the matrix element of Coulomb integral reads,

$$V_{prsq}^{e-h} = -\frac{e^2}{4\pi\epsilon_0\epsilon_r}\iint\frac{\chi_p^*(\mathbf{r}_e)\chi_r^*(\mathbf{r}_h)\chi_s(\mathbf{r}_h)\chi_q(\mathbf{r}_e)}{|\mathbf{r}_e-\mathbf{r}_h|}d\mathbf{r}_e d\mathbf{r}_h, \tag{55}$$

which involves the wave function χ_i of the electron and hole in the absence of Coulomb interaction, with $j = p, r, s, q$ indicating the state index. We should emphasize that Eq. 55 contains both the direct interaction between the electron and hole, i.e., Eq. 55 at $p = q$ and $r = s$,

$$V_{prrp}^{e-h,dir} = -\frac{e^2}{4\pi\epsilon_0\epsilon_r}\iint\frac{|\chi_p(\mathbf{r}_e)|^2|\chi_r(\mathbf{r}_h)|^2}{|\mathbf{r}_e-\mathbf{r}_h|}d\mathbf{r}_e d\mathbf{r}_h, \tag{56}$$

and the exchange interaction, i.e., Eq. 55 at $s = p$ and $r = q$,

$$V_{prpr}^{e-h,ex} = -\frac{e^2}{4\pi\epsilon_0\epsilon_r}\iint\frac{\chi_p^*(\mathbf{r}_e)\chi_r^*(\mathbf{r}_h)\chi_p(\mathbf{r}_h)\chi_r(\mathbf{r}_e)}{|\mathbf{r}_e-\mathbf{r}_h|}d\mathbf{r}_e d\mathbf{r}_h. \tag{57}$$

To calculate Coulomb interaction defined in Eq. 55, we expand $1/|\mathbf{r}_e - \mathbf{r}_h|$ in terms of half-integer Legendre function of the second kind $Q_{m-1/2}$, i.e.,

$$\frac{1}{|\mathbf{r}_e-\mathbf{r}_h|} = \frac{1}{\pi\sqrt{r_e r_h}}\sum_{m=0}^{\infty}\epsilon_m \cos\left[m(\theta_e-\theta_h)\right] Q_{m-1/2}(\xi), \tag{58}$$

which is widely used in many-body calculations [45]. Here $\theta_{e(h)}$ is the polar angle of the position vector $\mathbf{r}_{e(h)}$ in the 2D plane of MoS_2, $\xi = (r_e^2 + r_h^2)/2r_e r_h$, $\epsilon_m = 1$ for $m = 0$ and $\epsilon_m = 2$ for $m \neq 0$. The exciton energy and the corresponding wavefunction can be obtained by diagonalization of the matrix of many-particle Hamiltonian $H(\mathbf{r}_e, \mathbf{r}_h)$. With the exciton state at hand (Eq. 54), we are ready to the determine the excitonic absorption involving a transition from the ground state $|0\rangle$ to exciton state $|f\rangle = |\mathbf{\Psi}_{exc}\rangle$,

$$A(\omega) = \sum_f |\langle 0\mathcal{P}|f\rangle|\ \delta\{\hbar\omega - E_{exc}^{\nu}\}, \tag{59}$$

where E_{exc}^{ν} and $\hbar\omega$ are the exciton- and photon-energy, respectively, and $\mathcal{P} = \sum_{i,j}\delta_{sz,sz'}\langle\chi_i|\mathcal{H}_{L-M}|\chi_j\rangle\, a_{i,sz}\, h_{j,sz'}$ is the polarization operator, with $a_{i,sz}$ and $h_{j,sz'}$ being the electron and hole annihilation operators. After straightforward calculation, we finally obtain

$$A(\omega) = \sum_{\nu}\left(\delta_{sz,sz'}\sum_{i,j}C_{i,j}^{\nu}\langle\chi_i|\mathcal{H}_{L-M}|\chi_j\rangle\right)\delta\{\hbar\omega - E_{exc}^{\nu}\}. \tag{60}$$

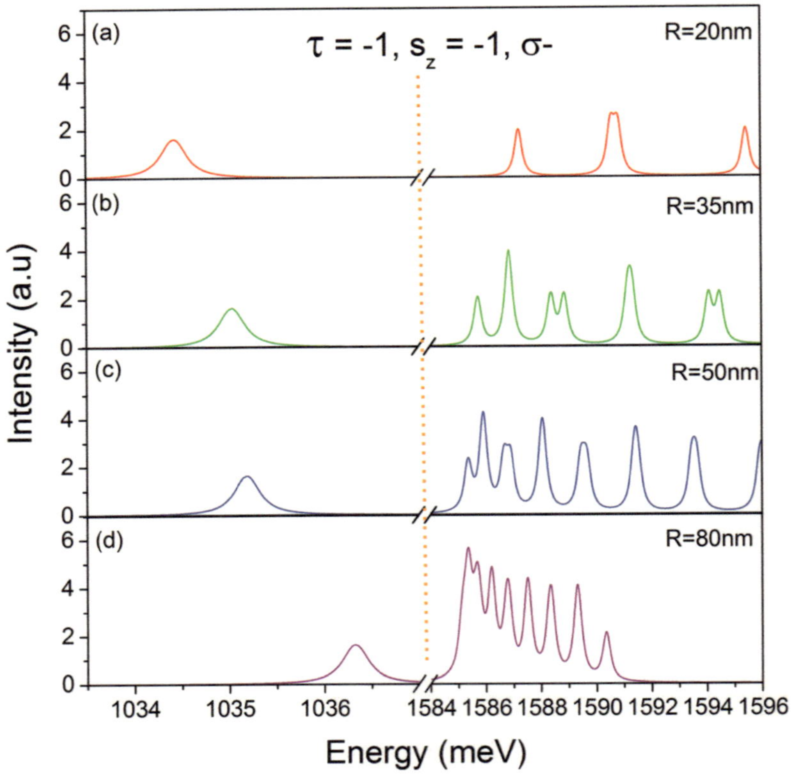

Figure 23. Excitonic effects on zero-field optical absorption spectra for the spin-down states in the K'-valley of monolayer MoS_2 QD with $R = 20$ nm (a), 35 nm (b), 50 nm (c), and 80 nm (d) under the excitation of the left-hand circularly polarized light σ^-. The Fermi energy is chosen as 0. The vertical dashed line separates two distinct regions of the absorption spectrum: left-hand side for excitonic absorption and right-hand side for single-particle like absorption. To be consistent with the experimental report of excitonic absorption energy, a right shift of photon energy of 700 meV is made [15]. (Copyright 2015 by the Nature Publishing Group. Reprinted with permission.)

In our numerical calculations, we have used five modified single-particle basis functions with angular momentum ranging from −2.5 to 1.5. **Figure 23** shows that there is an exciton absorption peak located at around 550 meV (i.e., exciton binding energy) below the band-edge absorption. And, the excitonic absorption peak shifts monotonically to higher absorption energy as the dot size is increased. Above the band gap, however, the spectrum is similar to what we found in previous sections in the band-to-band transitions using the independent electron-hole model. Since the exciton absorption peak is far away from the band-edge absorption, one can in principle study them separately. And, the Coulomb interaction between electron-hole pair does not change the valley selectivity and our general conclusion. Finally, it is worth to remarking that one can shift the excitonic absorption peak to a higher absorption energy, by varying the band gap parameter (Δ) in our model Hamiltonian.

Author details

Fanyao Qu[1]*, Alexandre Cavalheiro Dias[1], Antonio Luciano de Almeida Fonseca[1], Marco Cezar Barbosa Fernandes[1] and Xiangmu Kong[2]

*Address all correspondence to: fanyao@unb.br

1 Institute of Physics, University of Brasília, Brazil

2 Department of Physics, Qu Fu Normal University, China

References

[1] Reimann SM, Manninen M. Reviews of Modern Physics. 2002;**74**:1283

[2] Kuwenhoven LP, Austing DG, Tarucha S. Reports on Progress in Physics. 2001;**64**:701

[3] Heinrich AJ, Gupta JA, Lutz CP, Eigler DM. Science. 2004;**306**:466

[4] Qu F, Hawrylak P. Physical Reviews Letters. 2005;**95**:217206

[5] Qu F, Hawrylak P. Physical Reviews Letters. 2006;**96**:157201

[6] Besombes L, Léger Y, Maingault L, Ferrand D, Mariette H, Cibert J. Physical Reviews Letters. 2004;**93**:207403

[7] Léger Y, Besombes L, Maingault L, Ferrand D, Mariette H. Physical Reviews Letters. 2005;**95**:047403

[8] Gould C, Slobodskyy A, Supp D, Slobodskyy T, Grabs P, Hawrylak P, Qu F, Schmidt G, Molenkamp LW. Physical Reviews Letters. 2006;**97**:017202

[9] Villegas-Lelovsky L, Qu F, Massa LO, Lopez-Richard V, Marques GE. Physical Review B. 2011;**84**:075319

[10] Qu F, Villegas-Lelovsky L, Morais PC. Physical Review B. 2015;**92**:115445

[11] Li X, Rui M, Song J, Shen Z, Zeng H. Advanced Functional Materials. 2015;**25**:4929

[12] Srivastava A, Sidler M, Allain AV, Lembke DS, Kis A, Imamolu A. Nature Nanotechnology. 2015;**10**:491

[13] Pavlovi S, Peeters FM. Physical Review B. 2015;**91**:155410

[14] Dias AC, Fu J, Villegas-Lelovsky L, Qu F. Journal of Physics: Condensed Matter. 2016;**28**:375803

[15] Qu F, Dias AC, Fu J, Villegas-Lelovsky L, Azevedo DL. Scientific Reports. 2017;**7**:41044

[16] Castro Neto AH, Guinea F, Peres NMR, Novoselov KS, Geim AK. Reviews of Modern Physics. 2009;**81**:109

[17] Alessandro M, Joshua G, Houssa M, Xu Y, Zhang SC, Akinwande D. Journal of Nanomaterials. 2017;**16**:163

[18] Balendhran S, Walia S, Nili H, Sriram S, Bhaskaran M. Small 2015;**11**:633

[19] Xiao D, Liu GB, Feng W, Xu X, Yao W. Physical Reviews Letters. 2012;**108**:196802

[20] Song XX, Liu D, Mosallanejad V, You J, Han TY, Chen DT, Li HO, Cao G, Xiao M, Guo GC, Guo GP. Nanoscale. 2015;**7**:16867

[21] Chernikov A, van der Zande AM, Hill HM, Rigosi AF, Velauthapillai A, Hone J, Heinz TF. Physical Reviews Letters. 2015;**115**:126802

[22] Withers F, Del Pozo-Zamudio O, Mishchenko A, Rooney AP, Gholinia A, Watanabe K, Taniguchi T, Haigh SJ, Geim AK, Tartakovskii AI, Novoselov KS. Nature Materials. 2015; **14**:301

[23] Bellus MZ, Ceballos F, Chiu HY, Zhao H. ACS Nano. 2015;**9**:6459

[24] Mak KF, He K, Lee C, Lee GH, Hone J, Heinz TF, Shan J. Nature Materials. 2013;**12**:207

[25] Li Y, Qi Z, Liu M, Wang Y, Cheng X, Zhang G, Sheng L. Nanoscale. 2014;**6**:15248

[26] Lin JD, Han C, Wang F, Wang R, Xiang D, Qin S, Zhang XA, Wang L, Zhang H, Wee ATS, Chen W. ACS Nano. 2014;**8**:5323

[27] Lin Y, Ling X, Yu L, Huang S, Hsu AL, Lee YH, Kong J, Dresselhaus MS, Palacios T. Nano Letters. 2014;**14**:5569

[28] Zhang C, Wang H, Chan W, Manolatou C, Rana F. Physical Review B. 2014;**89**:205436

[29] Padilha JE, Peelaers H, Janotti A, Van de Walle CG. Physical Review B. 2014;**90**:205420

[30] Splendiani A, Sun L, Zhang Y, Li T, Kim J, Chim CY, Galli G, Wang F. Nano Letters. 2010; **10**:1271

[31] Heo H, Sung JH, Cha S, Jang BG, Kim JY, Jin G, Lee D, Ahn JH, Lee MJ, Shim JH, Choi HY, Jo MH. Nature Communications. 2015;**6**:7372

[32] Koperski M, Nogajewski K, Arora A, Cherkez V, Mallet P, Veuillen JY, Marcus J, Kossacki P, Potemski M. Nature Nanotechnology. 2015;**10**:503

[33] He YM, Clark G, Schaibley JR, He Y, Chen MC, Wei YJ, Ding X, Zhang Q, Yao W, Xu X, Lu JW, Lu CY, Pan JW. Nature Nanotechnology. 2015;**10**:497

[34] Sheng YB, Zhou L, Zhao SM, Zheng BY. Physical Review A. 2012;**85**:012307

[35] Gl AD, Potasz P, Hawrylak P. Graphene-Based Integrated Electronic, Photonic and Spintronic Circuitin Future Trends in Microelectronics: Frontiers and Innovations. Luryi S, Xu J, Zaslavsky A, editors. Wiley-IEEE Press; 2013. ISBN: 978-1-118-44216-6

[36] Wu F, Qu F, MacDonald AH. Physical Review B. 2015;**91**:075310

[37] Chakraborty C, Kinnischtzke L, Goodfellow KM, Beams R, Vamivakas AN. Nature Nanotechnology. 2015;**10**:507

[38] Lwdin PO. International Journal of Quantum Chemistry. 1982;**21**:69

[39] Liu GB, Shan WY, Yao Y, Yao W, Xiao D. Physical Review B. 2013;**88**:085433

[40] You Y, Zhang XX, Berkelbach TC, Hybertsen MS, Reichman DR, Heinz TF. Nature Physics. 2015;**11**:477

[41] Rose F, Goerbig MO, Piechon F. Physical Review B. 2013;**88**:125438

[42] Recher P, Trauzettel B, Rycerz A, Blanter YM, Beenakker CWJ, Morpurgo AF. Physical Review B. 2007;**76**:235404

[43] Mak KF, He K, Shan J, Heinz TF. Nature Nanotechnology. 2012;**7**:494

[44] Cao T, Wang G, Han W, Ye H, Zhu C, Shi J, Niu Q, Tan P, Wang E, Liu B, Feng J. Nature Communications. 2012;**3**:887

[45] Cohl HS, Rau ARP, Tohline JE, Browne DA, Cazes JE, Barnes EI. Physical Review A. 2001;**64**:052509

Chapter 8

Widely Tunable Quantum-Dot Source Around 3 μm

Alice Bernard, Marco Ravaro, Jean-Michel Gerard,
Michel Krakowski, Olivier Parillaud, Bruno Gérard,
Ivan Favero and Giuseppe Leo

Additional information is available at the end of the chapter

http://dx.doi.org/10.5772/intechopen.70753

Abstract

We propose a widely tunable parametric source in the 3 μm range, based on intracavity spontaneous parametric down conversion (SPDC) of a quantum-dot (QD) laser emitting at 1.55 μm into signal and idler modes around 3.11 μm. To compensate for material dispersion, we engineer the laser structure to emit in a higher-order transverse mode of the waveguide. The width of the latter is used as a degree of freedom to reach phase matching in narrow, deeply etched ridges, where the in-plane confinement of the QDs avoids non-radiative sidewall electron-hole recombination. Since this design depends critically on the knowledge of the refractive index of $In_{1-x}Ga_xAs_yP_{1-y}$ lattice matched to InP at wavelengths where no data are available in the literature, we have accurately determined them as a function of wavelength (λ = 1.55, 2.12 and 3 μm) and arsenic molar fraction (y = 0.55, 0.7 and 0.72) with a precision of $\pm 4 \times 10^{-3}$. A pair of dichroic dielectric mirrors on the waveguide facets is shown to result in a continuous-wave optical parametric oscillator (OPO), with a threshold around 60 mW. Emission is tunable over hundreds of nanometers and expected to achieve mW levels.

Keywords: quantum dots, laser diode, near infrared, InGaAsP, tunable source, OPO

1. Introduction

The tunability of currently available integrated sources is limited to a few tens of nanometers at most, via temperature or current control. While this is not a problem for most applications, certain fields like wavelength division multiplexing and spectroscopy are in demand of sources with broader tunability and choice of spectral range. Spectroscopy, especially, requires wide-band, continuously tunable sources with narrow emission lines. The 2–4 μm wavelength interval is of particular interest since it contains various peaks of atmospheric and hydrocarbon molecules, with important applications in environmental monitoring, security, and medicine [1–4]. This spectral region is at the frontier between the emission ranges of diodes and quantum cascade lasers (QCLs), and to date most existing sources around 3 μm, such as short-wavelength QCLs [5, 6] or GaSb diodes [7], are only available in laboratories. Interband cascade lasers (ICLs) are the only sources commercially available in this wavelength range, albeit at a high price [8–10]. Moreover, the tunability range of all these devices is limited to a few tens of nanometers. As a consequence, individual laser diodes are used for each spectroscopic line of absorption, which increases the price of a complete diagnosis based on several lines. In this context, nonlinear optics offers a solution for widely wavelength tunable sources, bulky tabletop optical parametric oscillators (OPOs) being commonly used to provide high-quality, tunable beams. Most miniaturized OPOs have been demonstrated in $LiNbO_3$ [11, 12], but an OPO threshold has been achieved in a GaAs micrometric waveguide [13] with a potential span of 500 nm. Like GaAs, InP is an attractive material for its high $\chi^{(2)}$ and mature technology, especially for emission at 1.55 μm.

Here we report on the design of an InGaAsP/InP QD laser diode emitting at 1.55 μm, optimized for intracavity spontaneous parametric down conversion (SPDC) around 3.11 μm via modal phase matching. The use of QDs is justified by the choice of narrow, deeply etched structures insofar they have been shown to trap carriers and limit surface recombination [14], and narrow-ridge, low-threshold InAs/InP QD lasers have already been demonstrated [15]. In order to estimate the phase mismatch accurately, a precise knowledge of the refractive indices is critical at pump, signal, and idler wavelengths. While the index of InGaAsP lattice matched to InP is well known at 1.55 μm [16–20], to date only one publication deals with its measurement at longer wavelengths [21], and none exists at 3 μm. This makes it crucial to accurately characterize its refractive index up to 3.14 μm, outside of the scope covered by literature data.

2. Tunable source design

2.1. Laser diode design

We propose a 1.55 μm source optimized for SPDC around 3.11 μm. This design results from back-and-forth optimizations between optical and electrical simulations, to jointly facilitate electron-hole injection, increase the conversion efficiency, and reduce losses. The conduction band and composition profile of this structure are shown in **Figure 1**. To compensate for the material dispersion, the laser diode is conceived so as to favor lasing on the TE_{20} mode (the second order in the direction of growth). To achieve this, the refractive index kept small in the center of the waveguide. As a consequence, the TE_{20} mode confinement inside the active area is stronger than for the TE_{00} mode. **Figure 2** shows the modes TE_{00} and TE_{20} supported by the waveguide at a wavelength of 1.55 μm. In order to achieve an efficient electron injection despite the conduction band increase in the core center, we reduce the series resistance with two strategies. Firstly, we introduce compositional gradients at the interfaces. Secondly, the waveguide core is only lightly doped. **Figure 3** depicts the conduction band and doping profile of the structure. An electrical simulation of this device using the software Nextnano yields a transparency current of 26 A/cm^2 at the transparency threshold.

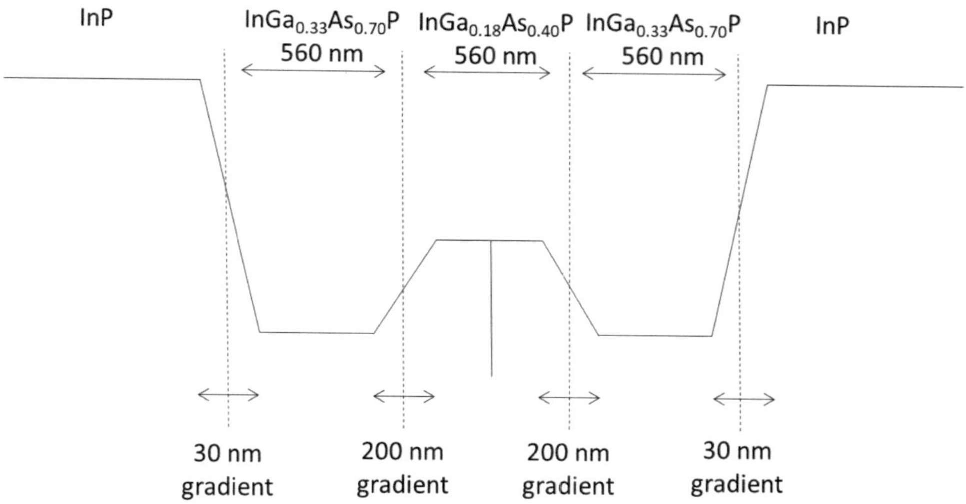

Figure 1. Conduction band and composition profile of the structure.

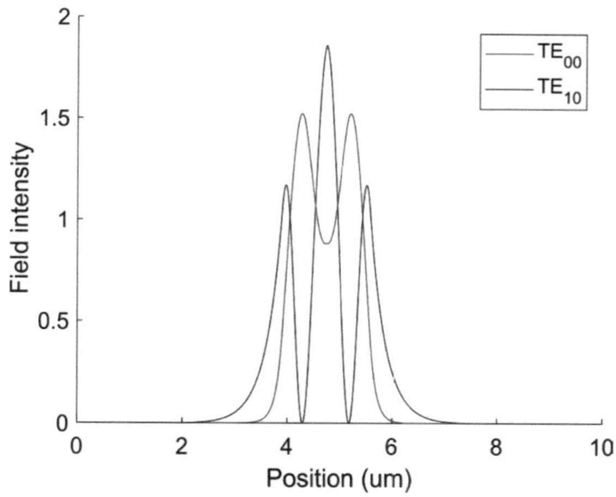

Figure 2. The first two even modes supported by the waveguide at $\lambda = 1.55$ μm.

2.2. Nonlinear properties

To achieve Type-II phase matching despite the error bars on the dispersion model and the fabrication tolerances, we use the ridge width as a crucial degree of freedom. **Figure 4** shows the phase mismatch at degeneracy vs. ridge width and pump wavelength, defined as

$$\Delta n = n\,(TE_{20}, 1.55\ \mu m) - [n\,(TE_{00}, 3.11\ \mu m) + n\,(TM_{00}, 3.11\ \mu m)]/2 \tag{1}$$

where the refractive indices are provided by our experimental data, presented in Part 2, and an interpolation of literature data [20]. By changing the ridge width from 3 to 7 μm, we are able to achieve phase matching for pump wavelengths of 1.50–1.60 μm. Furthermore, a variation in phase mismatch of ±0.02 can be compensated for by setting the correct ridge size. The ridge width thus acts as a gross parameter to meet the phase-matching condition, which can be set after wafers have been grown and characterized.

During operation, temperature provides a supplementary degree of freedom to tune the pump wavelength and reach a wide range of frequencies. **Figure 5** shows the wavelengths

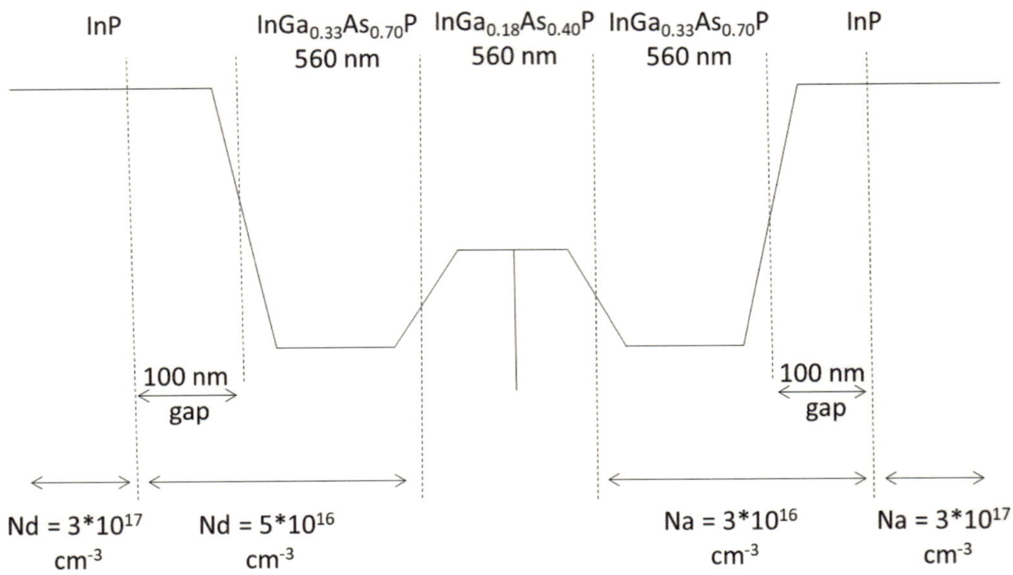

Figure 3. Conduction band and doping profile of the structure.

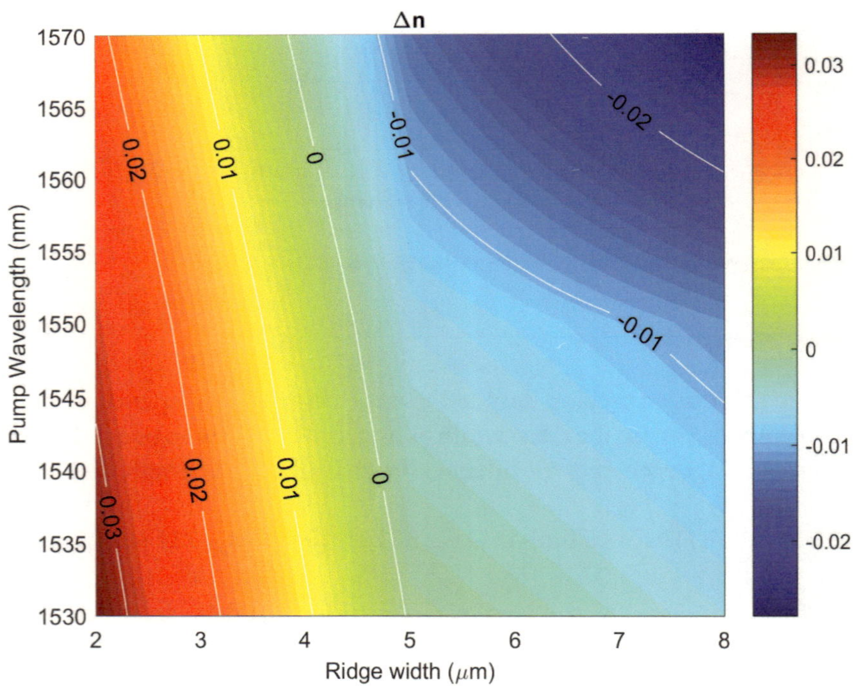

Figure 4. Phase mismatch in a deeply etched structure vs. ridge width and pump wavelength.

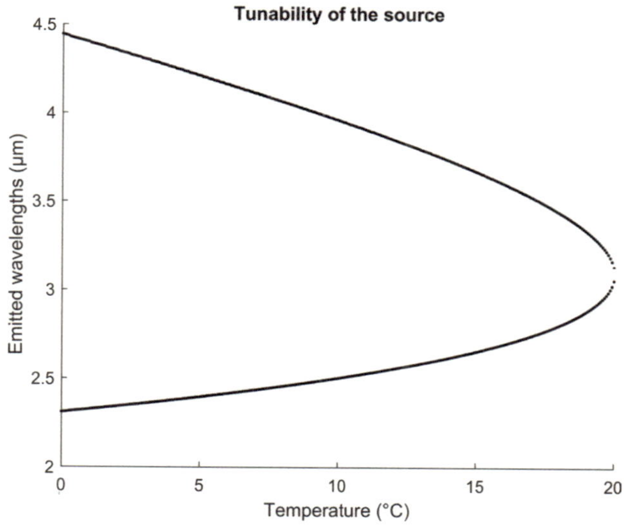

Figure 5. Emitted wavelengths for a device of ridge width 3.3 μm, emitting at 1.55 μm at 20°C.

of signal and idler emitted beams, for a source of ridge width 3.3 μm, emitting at a pump wavelength of 1.55 μm at 20°C. The dependency of quantum dots wavelength emission with temperature was assumed to be 0.5 nm/K from [22].

Figure 6 shows the profiles of the interacting modes, at a pump wavelength of 1.55 μm and signal and idler 3.11 μm. The expected conversion efficiency at a pump wavelength of 1.55 μm is 240% W^{-1} cm^{-2}. For an intracavity power of 100 mW, this corresponds to a parametric gain of 0.5 cm^{-1}. Since common InP QD lasers at 1.55 μm emit up to 20 mW outside the cavity without facet coatings [23], the above hypothesis on the intracavity power is very reasonable. The losses experimented by the mode at 3.11 μm are mainly expected to stem from free-carrier

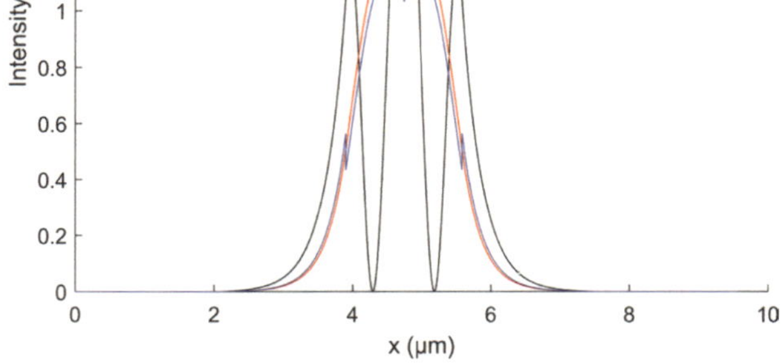

Figure 6. Field intensity of the three interacting modes in the SPDC process, at a pump wavelength of 1.55 μm.

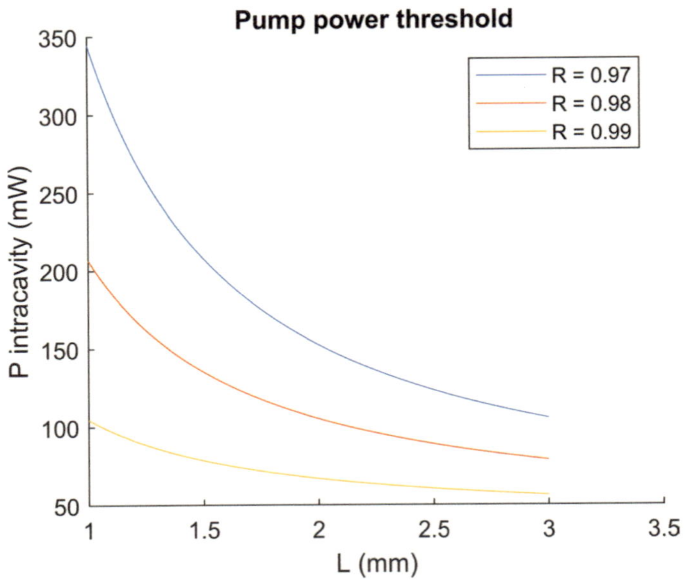

Figure 7. Intracavity pump power necessary for OPO threshold, vs. ridge length, for a few values of facet reflectivities.

absorption in doped layers, which we estimate at about 0.34 cm^{-1} for the TE_{00} mode. Reaching the OPO threshold in this device is therefore a challenging but achievable task, if mirror losses are minimized as in [13]. **Figure 7** shows the intracavity pump power necessary to reach oscillation for different mirror reflectivities and cavity lengths, assuming a conversion efficiency of 240%W^{-1} cm^{-2} and propagation losses of 0.34 cm^{-1}.

3. Index measurement

3.1. Principle

The evaluation of refractive indices was performed through an m-lines setup. This measurement relies on the determination of the coupling angle inside a slab waveguide of the material of interest. Our samples consist of a layer of $In_{1-x}Ga_xAs_yP_{1-y}$ lattice matched to InP grown over InP. As a result of the index contrast, the quaternary alloy acts as a planar waveguide. A diffraction grating was deposited or etched on the surface. The samples are mounted vertically on a rotating mount, and a laser is shone horizontally on the grating (**Figure 8**). Whenever the angle of light diffracted into the surface layer matches the bounce angle of a guided mode, coupling occurs. The guided light is collected by a detector placed at the exit facet of the sample. Coupling angles are given by the conservation of momentum

$$\left\|\overrightarrow{k_{in}}\right\| sin(\theta) = \left\|\overrightarrow{k_m}\right\| + m \left\|\overrightarrow{k_g}\right\| \quad (2)$$

where

$\|\vec{k}_{in}\|$ is the wave vector of the incident light, $\|\vec{k}_{m}\|$ is the wave vector of the mth guided mode, and $\|\vec{k}_{g}\|$ is the vector associated to the grating, of amplitude $2\pi/\Lambda$ (Λ being the period) and direction perpendicular to the grating lines, in the grating plane.

This equation leads to

$$N = sin(\theta) + m\ \lambda/\Lambda \tag{3}$$

where N is the effective index of the mth propagating mode. This measurement therefore estimates precisely the effective indices of the guided modes, as long as the period, wavelength, and coupling angles are known.

3.2. Sample fabrication

The planar structures were grown by molecular beam epitaxy and characterized by X-ray diffraction and photoluminescence prior to processing. An electronic resist with average thickness 70 nm was deposited on the surface (MAN2401 spin coated at 6000 rpm), and diffraction gratings were written by electronic lithography. The samples were then either used as such or very shallowly etched in a chemical solution before removing the resist. Both methods lead to shallow gratings on the guiding layer surface.

3.3. Optical setup

Three laser beams are combined with a precision of 0.01°: (1) a visible one (λ = 543 nm) for alignment and period measurements, (2) a tunable fibered laser around 1.55 μm, and (3) a free-space laser diode emitting at either 3.14 or 3.17 μm. The sample is placed on the axis of a rotating stage driven by a motor of step 0.001° and is precisely aligned so that the diffraction grating lines lie in the vertical direction (**Figure 9**). The light output at its exit facet is collected

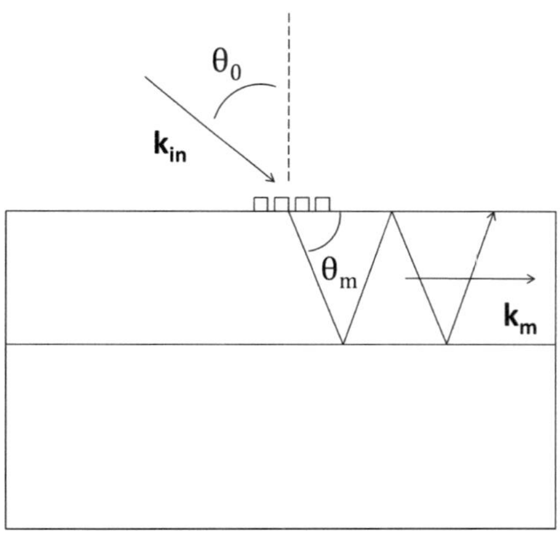

Figure 8. Coupling condition.

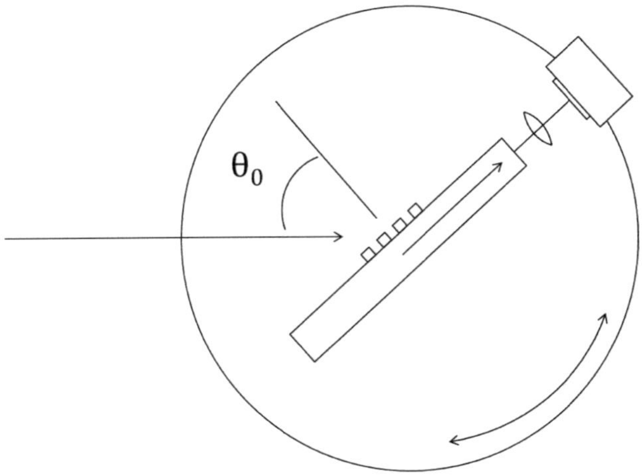

Figure 9. Optical setup for effective-index measurements.

with a detector placed after a slit (not shown here) and a focusing lens. All the detection setup is mounted on the stage, and it rotates rigidly with the sample. We used the same samples and gratings at 1.55 and 3 μm since the direction of the second-order diffracted beam at 1.55 μm roughly corresponds to the first order at 3.11 μm. To find the angle of normal incidence, measurements were done in a single long scan, with symmetric angles. Since all facets of a sample are cleaved and reflective, whenever θ_1 is a coupling angle, the light couples as well in the other direction for $\theta_2 = -\theta_1$ and reflects on the back facet, reaching the detector (**Figure 10**). Thus the angle of normal incidence was taken to be $(\theta_1 + \theta_2)/2$. This reduces the error bar due to a possible misalignment of the beams and to the motor backlash.

In order to determine the grating period, the visible laser was shone on the sample. Under the Littrow condition, the diffracted beam exits through the entrance slit and is collected through a beam splitter on a photodiode (**Figure 11**). The grating period is retrieved through the equality

$$2\,sin(\theta) = m\lambda\,/\Lambda \tag{4}$$

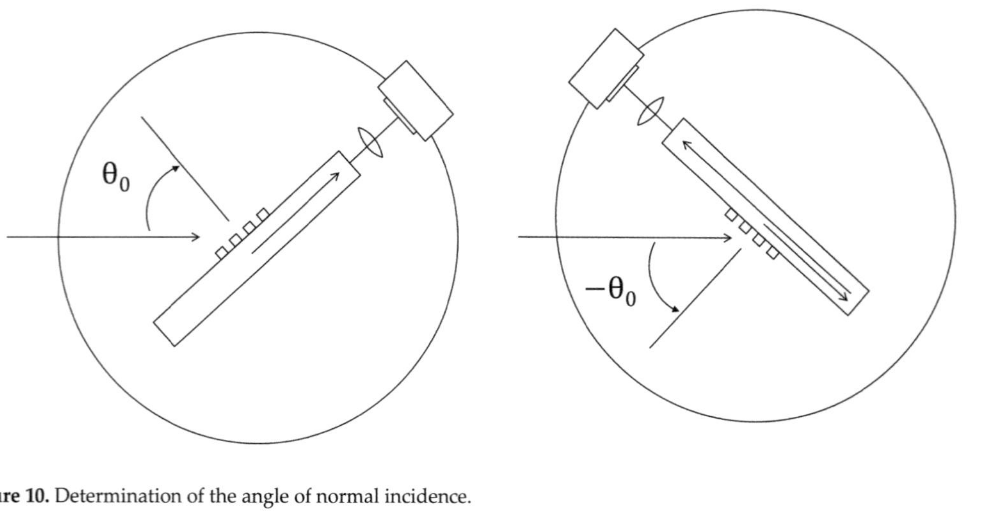

Figure 10. Determination of the angle of normal incidence.

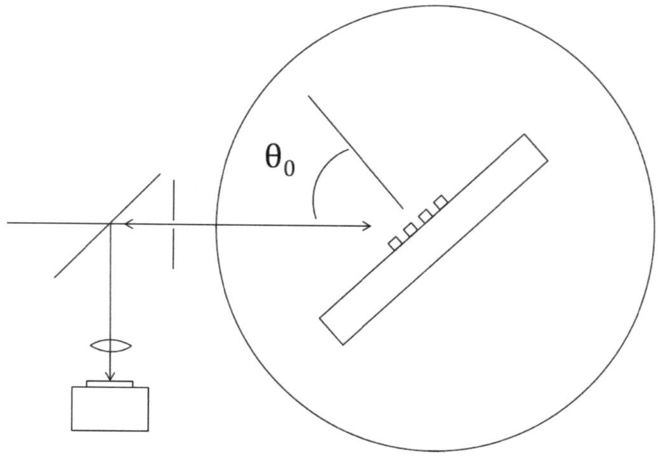

Figure 11. Optical setup for measuring the grating period.

3.4. Results

A typical result of coupling measurement is shown in **Figure 12**. The position of the peaks is determined with a precision of 0.01°. Since the effective indices are a function of material index and thickness of the guiding layer, each measured value corresponds to a range of possible {material index, thickness} pairs. This is represented in **Figure 13**, where one line corresponds to the space of parameters that minimize the difference between measured and theoretical indexes. Since more than one effective index is measured, it is possible to determine the right pair {material index, thickness}, at the crossing point. **Figure 14** shows the average difference between measured and effective indices. Waveguides support three modes at 1.55 μm, two at

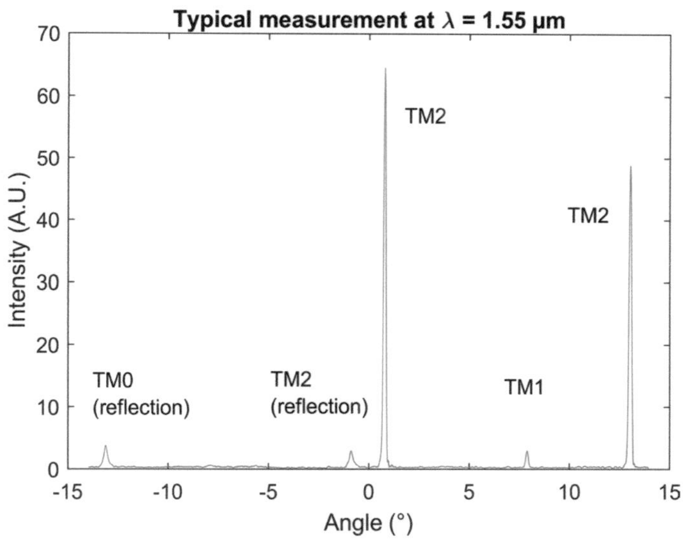

Figure 12. Determination of the refractive index and thickness for a slab of $In_{0.67}Ga_{0.33}As_{0.72}P_{0.28}$ at λ = 1.55 μm. Each line shows the possible range of data corresponding to the measured value of the effective index of a given waveguide mode.

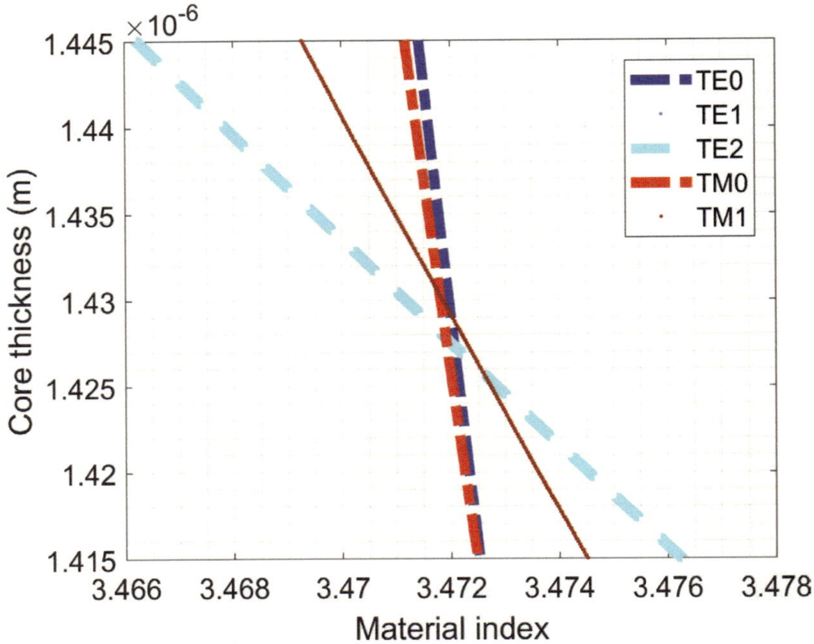

Figure 13. Coupling measurement into a slab of $In_{0.67}Ga_{0.33}As_{0.72}P_{0.28}$ on InP at a wavelength of 1.55 μm.

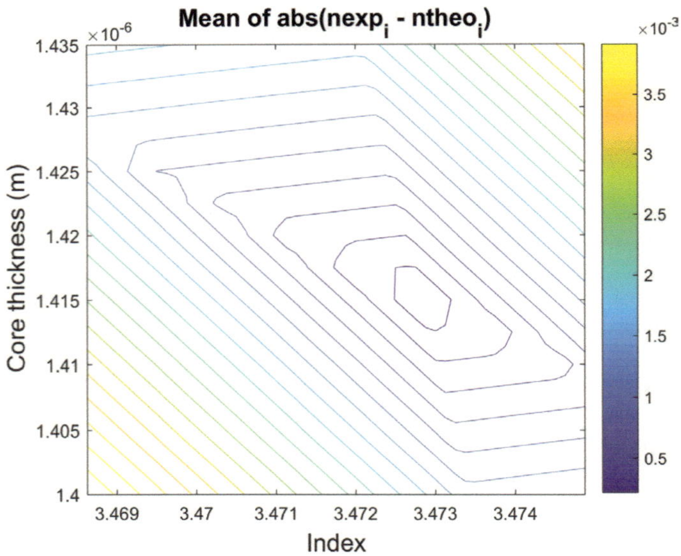

Figure 14. Determination of the refractive index and core thickness for a slab of $In_{0.67}Ga_{0.33}As_{0.72}P_{0.28}$ on InP at a wavelength of 1.55 μm. This figure shows the mean difference between calculated and measured effective indices. Area width gives an estimation of the error.

2.12 μm, and one at 3.14 μm. In order to derive the effective index at 3 μm, the guide thicknesses were estimated from the data at 1.55 μm (**Table 1**).

Figures 15–17 show the measured refractive indices as a function of As fraction, at wavelengths of 1.55, 2.12, and 3.14 μm. The data at 1.55 μm was compared to the model presented in [20]. At 2.12 and 3.14 μm, no model being available in the literature, we trace the data against a linear regression versus the molar fraction of As(y). The refractive index of InP from [24] was taken into account. **Figure 18** shows the refractive index versus wavelength, against a one-oscillator fit calculated from the Afromowitz model [25]. The fit parameters are presented in **Table 2**.

#	PL peak (μm)	da/a0	xGa	yAs	N (λ = 1.55 μm)	N (λ = 2.1 μm)	N (λ = 3.11 μm)
1	1.395	−0.0025	0.33	0.72	3.470		3.348
2	1.395	+0.002	0.33	0.72	3.472		3.349
3	1.395	n.m.	0.33	0.72		3.384	
4	1.346	n.m.	0.35	0.70	3.445		3.333
5	1.346	−0.0017	0.35	0.70	3.444		3.330
6	1.346	n.m.	0.35	0.70		3.370	
7	1.266	n.m.	0.24	0.55	3.391		3.289
8	1.266	+0.00078	0.24	0.55		3.325	

The lattice mismatch was measured by X-ray diffraction. Ga and As fraction are deduced through the model presented in [26].

Table 1. Physical properties and measured indices of the studied samples.

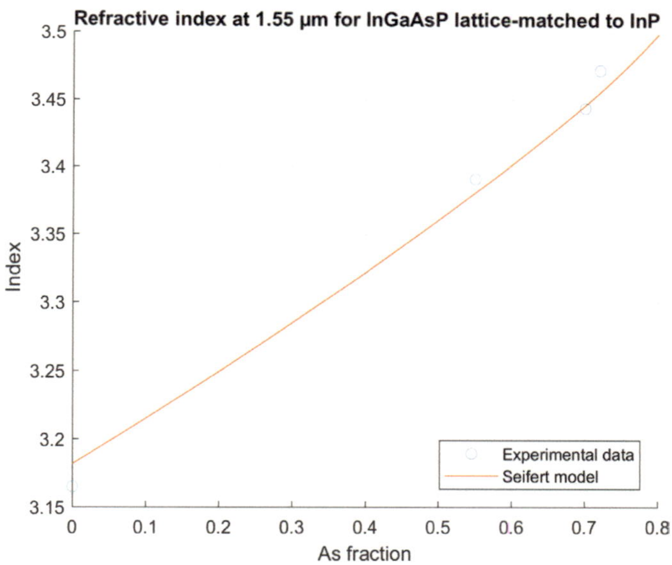

Figure 15. Refractive index measured at 1.55 μm vs. y, compared to [20].

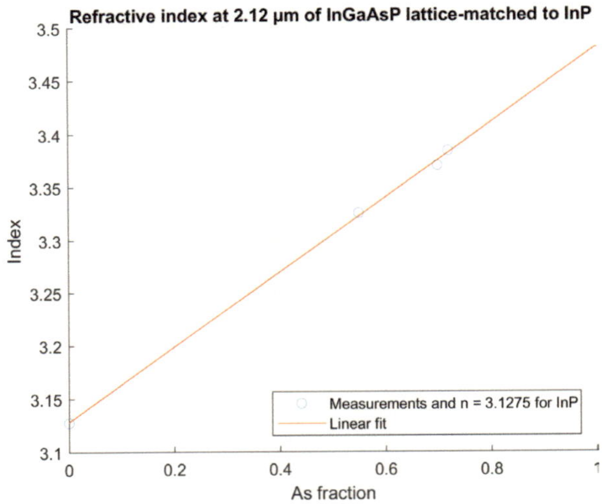

Figure 16. Refractive index measured at 2.12 μm vs. y (data plotted against a linear fit).

3.5. Discussion

The accuracy of this measurement is determined by several factors.

- Values of the grating period were determined by repeated measurements with a precision of 0.1 nm. This leads to a 0.3×10^{-3} error bar on the effective index.
- The uncertainty due to sample misalignment can be estimated at 0.5×10^{-3}. This may be explained by local variations of the resist filling factor and depth.
- In order to estimate the impact of the thin layer of photoresist on the effective indices, we performed a set of measurements on a sample covered with a thin photoresist grating. Then we etched it shallowly, removed the resist, and took a new set of data. The estimated thickness diminishes by 11 nm. This is in agreement with a profilometry of the etched grating depth, yielding 15 nm. The estimated core index is raised by 0.7×10^{-3}, a value lower than the possible variation of twice the experimental error. Thus we conclude that the resist has a negligible impact on the effective indices.
- While the laser beam alignment and position of the sample with respect to the rotating stage are adjusted in each measurement, one could point out that the axis of the rotating motor could be slightly misaligned with respect to the vertical axis and introduce a systematic error. A simple observation of the height of the beam reflection as the motor rotates indicates that the angle could be at most of 0.5 mrad. This leads, after a calculation, to an error on the effective index of 3×10^{-5}.
- Finally, incertitude on the composition is the most important. It is determined by the photoluminescence and lattice mismatch of the samples, with a precision of ~1%, through the model described in [26]. This corresponds to an uncertainty on the effective index of 4×10^{-3}. The deviation of our measurements with respect to literature and to a linear fit is in the range of 10^{-2} to 2×10^{-2}. This is in agreement with the observed variations of index due to the lattice mismatch observed in [16].

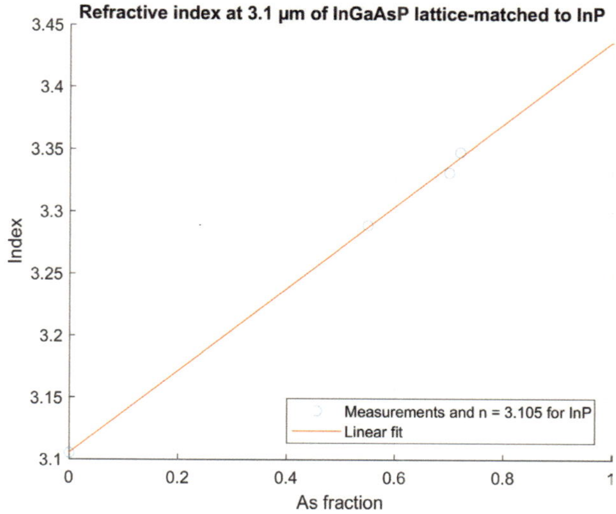

Figure 17. Refractive index measured at 3.14 μm vs. y (data plotted against a linear fit).

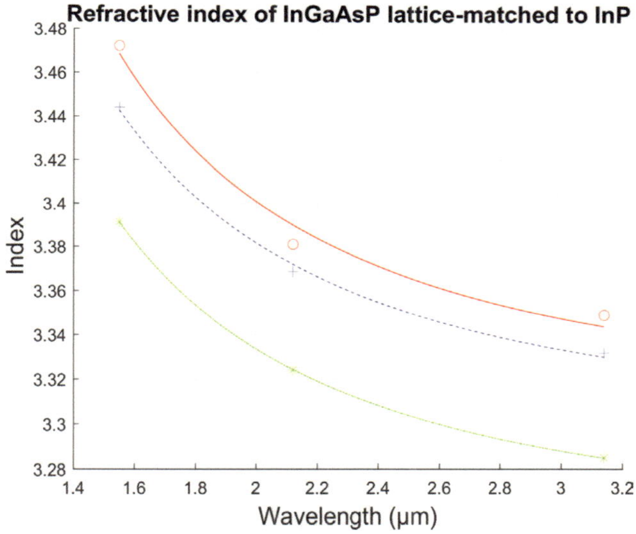

Figure 18. Refractive index of InGaAsP lattice matched to InP vs. wavelength for y = 0.72, 0.70, and 0.55. A one-oscillator fit (Afromowitz model) is added.

yAs	a	b
0.72	−0.0156	0.1007
0.70	−0.0144	0.1014
0.55	−0.0143	0.1044

Table 2. Parameters of the Afromowitz model inferred from the index measurements: a and b are extracted by a linear regression from $(n^2-1)^{-1} = a\,E^2 + b$, where E is the wavelength energy in eV.

Acknowledgements

This work is supported by a public grant overseen by the French National Research Agency (ANR) as part of the project DOPO. Authors thank the Commissariat à l'Energie Atomique and Direction Générale de l'Armement for PhD funding.

Author details

Alice Bernard[1,2], Marco Ravaro[1], Jean-Michel Gerard[2], Michel Krakowski[3], Olivier Parillaud[3], Bruno Gérard[3], Ivan Favero[1] and Giuseppe Leo[1*]

*Address all correspondence to: giuseppe.leo@univ-paris-diderot.fr

1 Laboratoire Matériaux et Phénomènes Quantiques, UMR 7162, Université Paris Diderot – CNRS, Paris, France

2 University Grenoble Alpes, CEA, INAC-PHELIQS, Grenoble, France

3 III-V Lab, Thales Research and Technology, Palaiseau, France

References

[1] Scholle K, Lamrini S, Koopmann P, Fuhrberg P. Bishnu Pal. 2 μm laser sources and their possible applications. In: Frontiers in Guided Wave Optics and Optoelectronics. InTech; 2010. ISBN 978-953-7619-82-4. February 1, 2010 under CC BY-NC-SA 3.0 license

[2] Siebenaler SP, Janka AM, Lyon D, Edlebeck JP, Nowlan AE. Methane detectors challenge: Low-cost continuous emissions monitoring. In: 2016 11th International Pipeline Conference 2016 Sep 26. American Society of Mechanical Engineers. New York, United States of America, ISBN : 978-0-7918-5027-5 pp. V003T04A013-V003T04A013

[3] Frish MB. Current and emerging laser sensors for greenhouse gas sensing and leak detection. In: Proceedings of the SPIE Sensing, Technologies & Applications Conference. Baltimore, MD, USA: Next Generation Spectroscopic Technologies VII; 21 May 2014. pp. 5-9

[4] Wang C, Sahay P. Breath analysis using laser spectroscopic techniques: Breath biomarkers, spectral fingerprints, and detection limits. Sensors. 2009 Oct 19;**9**(10):8230-8262

[5] Bandyopadhyay N, Bai Y, Tsao S, Nida S, Slivken S, Razeghi M. Room temperature continuous wave operation of $\lambda \sim$ 3-3.2 μm quantum cascade lasers. Applied Physics Letters. 2012 Dec 10;**101**(24):241110

[6] Devenson J, Barate D, Cathabard O, Teissier R, Baranov AN. Very short wavelength ($\lambda = 3.1 - 3.3$ μm) quantum cascade lasers. Applied Physics Letters. 2006 Nov 6;**89**(19): 191115

[7] Gaimard Q, Cerutti L, Teissier R, Vicet A. Distributed feedback GaSb based laser diodes with buried grating. Applied Physics Letters. 2014 Apr 21;**104**(16):161111

[8] Dong L, Li C, Sanchez NP, Gluszek AK, Griffin RJ, Tittel FK. Compact, low power consumption methane sensor based on a novel miniature multipass gas cell and a CW, room temperature interband cascade laser emitting at 3.3 μm. In: SPIE OPTO 2016 Feb 13. International Society for Optics and Photonics. Bellingham, Washington USA. pp. 97550I-97550I

[9] Manfred KM, Ritchie GA, Lang N, Röpcke J, van Helden JH. Optical feedback cavity-enhanced absorption spectroscopy with a 3.24 μm interband cascade laser. Applied Physics Letters. 2015 Jun 1;**106**(22):221106.

[10] Ren W, Luo L, Tittel FK. Sensitive detection of formaldehyde using an interband cascade laser near 3.6 μm. Sensors and Actuators B: Chemical. 2015 Dec 31;**221**:1062-1068

[11] Sohler W, Hu H, Ricken R, Quiring V, Vannahme C, Herrmann H, Büchter D, Reza S, Grundkötter W, Orlov S, Suche H. Integrated optical devices in lithium niobate. Optics and Photonics News. 2008 Jan 1;**19**(1):24-31

[12] Pomarico E, Sanguinetti B, Gisin N, Thew R, Zbinden H, Schreiber G, Thomas A, Sohler W. Waveguide-based OPO source of entangled photon pairs. New Journal of Physics. 2009 Nov 24;**11**(11):113042

[13] Savanier M, Ozanam C, Lanco L, Lafosse X, Andronico A, Favero I, Ducci S, Leo G. Near-infrared optical parametric oscillator in a III-V semiconductor waveguide. Applied Physics Letters. 2013 Dec 23;**103**(26):261105

[14] Moore SA, O'Faolain L, Cataluna MA, Flynn MB, Kotlyar MV, Krauss TF. Reduced surface sidewall recombination and diffusion in quantum-dot lasers. IEEE Photonics Technology Letters. 2006 Sep;**18**(17):1861-1863

[15] Barbarin Y, Anantathanasarn S, Bente EA, Oei YS, Smit MK, Notzel R. 1.55 μm range InAs–InP (100) quantum-dot Fabry–Pérot and ring lasers using narrow deeply etched ridge waveguides. IEEE Photonics Technology Letters. 2006 Dec 15;**18**(24):2644-2646

[16] Broberg B, Lindgren S. Refractive index of $In_{1-x}Ga_xAs_yP_{1-y}$ layers and InP in the transparent wavelength region. Journal of Applied Physics. 1 May 1984;**55**(9):3376-3381

[17] Amiotti M, Landgren G. Ellipsometric determination of thickness and refractive index at 1.3, 1.55, and 1.7 μm for $In_{1-x}Ga_xAs_yP_{1-y}$ films on InP. Journal of Applied Physics. 1993 Mar 15;**73**(6):2965-2971

[18] Henry C, Johnson L, Logan R, Clarke D. Determination of the refractive index of InGaAsP epitaxial layers by mode line luminescence spectroscopy. IEEE Journal of Quantum Electronics. 1985 Dec;**21**(12):1887-1892

[19] Fiedler F, Schlachetzki A. Optical parameters of InP-based waveguides. Solid State Electronics. 1987 Jan 31;**30**(1):73-83

[20] Seifert S, Runge P. Revised refractive index and absorption of $In_{1-x}Ga_xAs_yP_{1-y}$ lattice-matched to InP in transparent and absorption IR-region. Optical Materials Express. 2016 Feb 1;**6**(2):629-639

[21] Chandra P, Coldren LA, Strege KE. Refractive index data from $In_{1-x}Ga_xAs_yP_{1-y}$ films. Electronics Letters. 1981 Jan 8;**17**(1):6-7

[22] Landin L, Borgström M, Kleverman M, Pistol ME, Samuelson L, Seifert W, Zhang XH. Optical investigation of In:As/InP quantum dots at different temperatures and under electric field. Thin Solid Films. 2000 Mar 27;**364**(1):161-164

[23] Reithmaier JP, Eisenstein G, Forchel A. InAs/InP quantum-dash lasers and amplifiers. Proceedings of the IEEE. 2007 Sep;**95**(9):1779-1790

[24] Pettit GD, Turner WJ. Refractive index of InP. Journal of Applied Physics. 1965 Jun 1; **36**(6):2081

[25] Afromowitz MA. Refractive index of $Ga_{1-x}Al_xAs$. Solid State Communications. 1974 Jul 1;**15**(1):59-63

[26] Kuphal E. Phase diagrams of InGaAsP, InGaAs and InP lattice-matched to (100) InP. Journal of Crystal Growth. 1984 Aug 1;**67**(3):441-457